PROFESSIONAL
MODELMAKING

D1372618

PROFESSIONAL MODELMAKING

A HANDBOOK OF TECHNIQUES
AND MATERIALS FOR
ARCHITECTS AND DESIGNERS

NORMAN TRUDEAU

WHITNEY LIBRARY OF DESIGN
an imprint of Watson-Guptill Publications/New York

FOR BERNADETTE

Art on page 2: Telephone. Photo, design and model: Shu Takahama.

Art on page 142 reproduced from The Measure of Man and Woman, *Henry Dreyfuss Associates.*

Senior Editor: Roberto de Alba
Associate Editor: Micaela Porta
Designer: Areta Buk
Production Manager: Hector Campbell

Copyright © 1995 by Norman Trudeau

First published in the United States in 1995 by Whitney Library of Design,
an imprint of Watson-Guptill Publications, a division of BPI Communications, Inc.,
1515 Broadway, New York, NY 10036.

Library of Congress Cataloging-in-Publication Data

Trudeau, Norman, 1939–
    Professional modelmaking : a handbook of techniques and materials
    for architects and designers.
        p.   cm.
    ISBN 0-8230-4098-4 : $35.00
    1. Engineering models.   2. Models and modelmaking.
    3. Architectural models.   I. Title.
    TA177. T78   1995
    745.2'022—dc20   94-42679
                                                        CIP

Distributed in the United Kingdom by Phaidon Press, Ltd., 140 Kensington
Church Street, London W8 48N, England.

Distributed in Europe (except the United Kingdom, South and Central America,
the Caribbean, the Far East, the Southeast, and Central Asia) by Rotovision S.A.,
Route Suisse 9, CH-1295 Mies, Switzerland.

Manufactured in Mexico
First printing, 1995

1 2 3 4 5 6 7 8 9 / 02 01 00 99 98 97 96 95

CONTENTS

CHAPTER 5

FOAM CORE

CHAPTER 6

FOAMS

CHAPTER 7

THERMOPLASTICS AND THERMOSETS

CHAPTER 8

STYLING CLAY

CHAPTER 9

SOLVENT BONDING

CHAPTER 10

GLUES AND CEMENTS

INTRODUCTION

The result of nearly thirty years of experience in modelmaking, free-lance design and industrial production, this book is a complete guide to modelmaking for designers and architects, as well as for students and amateurs. Over a hundred new and old materials used in modelmaking are described, along with the techniques and tricks modelmakers of all levels should know to use them effectively. Since a number of materials commonly used in modelmaking are harmful to humans and the environment, strong emphasis has been placed on health, safety and ecological concerns in the hope that such issues will someday become important priorities in the study and practice of industrial design.

The model, or maquette, is a study or prototype of a three-dimensional design project. Viewed and understood more easily than sketches or drawings, the model photographs better and is more striking and effective in design presentations because it simply comes closer to the final product in mind. If well made, the model can look as "real" as the projected outcome.

In fields such as art, architecture, archaeological restoration, television, theatre and motion picture set design, there is an increased desire to work with all the materials and techniques available today. And while some designers may never lay hands on a tool, they may benefit from the practical information contained in this book as much as modelmakers. The technical explanations presented here drastically reduce the need for the very guesswork which often leads to undesirable results. A designer or design manager may also have to employ and judge the work of a modelmaker. Even in secondary schools, where manual skills and crafts teaching have always been limited by the exclusive use of wood, metal and paper, there is a yearning to expand beyond traditional boundaries and experiment with new materials.

Before reading this book, keep in mind that none of this material need be memorized. The information may be skimmed through, referred to back and forth from one subject to another or even read backwards, as some people read magazines. The important thing is that you be able to access the necessary information quickly, as it is needed. Practice will, of course, take care of the rest. Remember, too, that behind every modelmaking accident—as well as success—lies another learning opportunity.

I wish to offer many thanks to my former students, who are now professionals all over the world. Their photographed work and their enthusiasm have made this book a pleasure to write.

AUTHOR'S NOTE:
SAFETY, ECOLOGY AND
INDUSTRIAL HYGIENE

Much is said in this book about the specific risks of using modelmaking chemicals. As designers we are in a good position to do something about the environmental deterioration of our planet, and—equally important—the occupational risks to our health, namely by making informed decisions and choosing our materials wisely.

The use and presence of many chemicals in our workplaces cause illness and death in people, animals and plants. The relationship between such chemicals and the development of cancer, brain damage, genetic mutations, nervous disorders, skin ailments and vague "sensitivities" is only sometimes acknowledged by industry managements. All too often, risks are ignored in the name of cost-effectiveness and convenience. In the United States it took years before the coal, stone, wood and flour industries, politicians and even medical authorities admitted that diseases such as silicosis are dust-produced, despite the fact that their European counterparts already possessed documented proof.

In modelmaking, the "sensitivities" acquired through working with chemicals such as epoxies are permanent, requiring lifelong changes in working conditions. Medical and occupational compensation doesn't exist because direct harm has not yet been proven outside manufacturers' research facilities. The same manufacturers claim in their brochures that the reactivity of their products makes them too useful to change (the same reactivity that makes them so harmful to human cells) and that the products are perfectly safe, as long as the safety procedures outlined are followed. In most cases, those procedures are so impractical that everyone in the business knows that no one will ever follow them. Long exposure to acetone, methyl chloride, toluene, Polyvinyl alcohol (PVA), styrene vapors, resins or catalysts (the list is long) is felt in the form of a constant ache in the back of the head. Even after weeks away from the studio spent in a cleaner atmosphere, you not only still feel the achiness, but can smell and taste the chemicals as well. Clearly, these are symptoms of much larger problems.

My intention here is not to scare the reader, but rather to make him/her aware that design modelmaking uses many potentially harmful products. Anyone working with them should be well acquainted with their possible side effects, and so should proceed both cautiously and responsibly. There are many teachers and managers who knowingly throw poisonous chemicals into a sink drain and allow students and employees to take risks while using poisonous substances. On a typical clean-up day at the end of a modelmaking term in design schools, several big truckloads of waste must be carted away. This waste may contain up to a hundred different chemical products, some in need of high-temperature incineration or specific chemical reactions to render them safe, others simply untreatable. Mishandled by being disposed of together in the same dump, the waste eventually seeps into and contaminates drinking water supplies.

At this point the best thing we can do is recognize that safe options do exist, and implement them whenever possible. Plaster work is messy and can be slow, but it has proven it's safety over centuries of work in modelmaking. Plaster makes good models, and may be used for shaping, molding and casting work, or as a base for safe fillers. Other casting and molding materials like gelatin, wax and sand have also proven both their safety and effectiveness over time. Pottery clay (just plain earth) has been sculpted, cast and wrought for thousands of years, producing beautiful work. This material is so safe that it is actually used in skin care. It also requires few, easily learned techniques long known to artists. In addition to these time-tested materials, there are some new products on the market which are very safe to use, among them the flexible dental molding compounds that are fast, precise and inexpensive. The list of safe, proven materials could be very long, and many are listed in the appendix at the end of this book.

Right: Styling clay's smooth finish makes it a popular choice for car models. Photo, design and model: Shu Takahama.

But a book such as this cannot possibly cover all one needs to know regarding personal safety and environmental responsibility in our industry. Begin by setting limits, by deciding that certain chemicals are just not worth the risk to you or others, and then seek out alternatives. Insist on the safe use and proper disposal of dangerous or questionable chemicals that you do decide to continue using. And when dealing with suppliers, make sure to get as much information as possible on the materials you are using; you may even try to learn some very basic chemistry to help you in this endeavor. The following chapters present a sound basis for any type of modelmaking work with an emphasis on simple technical information and safety.

Photo, design and model: Marcel Soiron.

DESIGN MODEL
CRITERIA

WHAT THE DESIGN MODEL CAN BE

The role of the design model is sometimes misunderstood. In architecture and product, transport and packaging design, the model can be considered a nuisance for reasons ranging from cost to the effort expended in construction. Furthermore, there are designers whose knowledge of modelmaking materials and techniques is so deficient that they invariably fail to see the model's importance in design presentations. But as the strong interest of industrial clients and young designers in models shows, there is really no more effective a way of presenting three-dimensional design concepts.

Probably the best way to begin is by establishing precisely what a model is not. A model for design is not necessarily a prototype, and in that sense does not even have to be a "working" model—that is, one with moving parts. It is not a master model, an original used as a point of reference for final dimensions needed for eventual production. Nor is it a lasting construction, destined for interminable storage, endless shipping or momentous presentation to future grandchildren.

A well-made model is very often just what it takes to sell a design concept. Whether a designer is briefing his manager and colleagues or presenting to clients, the audience's interest will almost immediately focus on the model presented, regardless of their area of expertise. Many design concepts are confusing, even when presented well in drawings. Design models have the advantage of offering an immediately accessible understanding of the proposal, making them that much more useful and effective.

Even a mock-up, or sketch model (when time and budget are limited) is often better than flat work. Photographs of design events in the press, publicity shots and formal presentations concentrate on the model's real power to show, rather than on graphic work, no matter how good it may be. Models simply photograph better, and if they are well made, can look very similar to the final product proposed.

Fig. 1-1. A well-made model makes for a strong presentation. Photo, design and model: Ulrich Schraudolf.

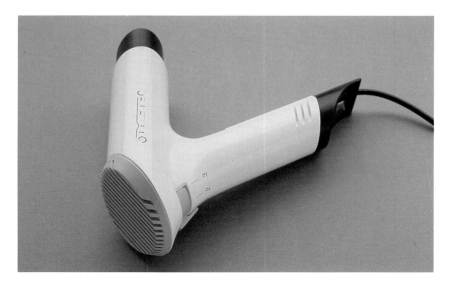

MODEL TYPES AND THEIR PURPOSES

Going further than a mock-up, the model can be a "finished" or "hard" model. This may be quite demanding, not only in the time and expense of actual construction but also in the time and expense of developing the facilities and training needed for good modelmaking. Some acceptable, but limited, model work can be done with a few simple tools and little expertise. On the other hand, some of the most experienced craftsmen will admit that they have never stopped learning and improving on their work. Model shops, too, are never quite up-to-date in their inventory because of the sheer multitude of materials and techniques available today.

A design model may be the "first word" in a design project, preceding even sketches or drawings. Using the appropriate materials and techniques, it can be built to allow for easy design changes, thereby saving time that would have been spent revising drawings. This may often happen after critical presentations, and there is usually no problem in adding to or subtracting from a model. Some designers and designs seem to lend themselves easily to this "reverse" approach, getting right into the three dimensions of the idea. Once the model is made and the design accepted, drawings and even technical, dimensioned work for future production and reference can be executed. This approach is fairly new to the industry, however, and so has not yet taken hold.

A model's purpose should also be paramount in its construction. If it is to be shipped from place to place, for example, fragile materials should not be used in its construction. If it is to be presented in brochure photographs, omitting some of the finer detail and playing with light and distance might make a better image. If formal presentations, close-up viewing and inevitable touching are expected, it should be the best model that your budget and time will allow. In a composite model, made to represent moving parts, the parts can be made and presented separately in their intended configurations.

There is already some discussion as to the future of the design model. Three-dimensional computer-assisted graphics have convinced some designers that modelmaking is now limited. In fact, the opposite is true. Despite certain kinds of work, like some architectural studies where computer-assisted graphics can save time, limitations exist in both computer representations and computer-assisted production of models for design. The actual models appear crude, and the programs for 3-D design may always be limited since computer programmers may not be primarily interested in design. Such innovations seem geared to please designers who need to play with a new toy—the computer—or those who know little about modelmaking, or so little about design itself that every curve can be measured with a radius gauge. There is, in fact, no limit to the number of forms that can be created with models.

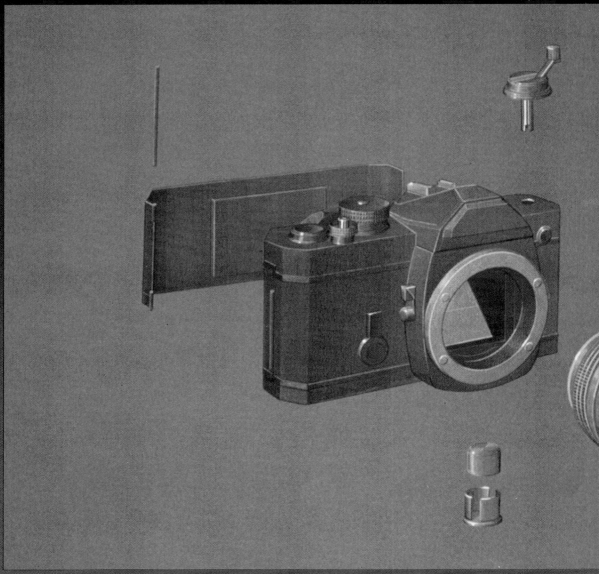

Drawing by William Wright.

CHAPTER 2

THREE-
DIMENSIONAL
FORM
DEVELOPMENT

DESIGNING WITH
DRAWINGS AND MODELS

The relationship between drawings and models is important not only in design, but in all industries. Both the drawing and the model serve specific functions. The drawing is generally used:

- As the initial two-dimensional vehicle for the design study.
- To guide the production of a model, prototype or finished product to dimensions and specifications.
- For client presentation of design concepts.
- As documentation for future reference (with revisions noted, for example) needed for production, repair, client service, etc..

The final drawing should be as clear as possible (but without unnecessary details) to allow for a quick appreciation of what it represents. Nevertheless, there still remain limits as to what a drawing can quickly and clearly show. The model, in three-dimensional design, also has its usefulness:

- It can give a more immediate appreciation, or "feeling," of the product than a drawing attempts to communicate.
- It may be the only way to seriously study highlights (light reflection), surface textures, finishes, etc..
- It is necessary for such studies as interior configuration and comfort.

GOING FORWARD OR BACKWARD

There are two basic approaches to designing in three dimensions. The first is to produce sketches, drawings and renderings which are studied for design decisions. The drawings are then dimensioned and used to guide the production of a model or prototype. This is most often done, both in design schools and in the "real" world of industrial design.

The second approach is more rare: designing by first building a model, studying it and incorporating design changes directly into the model. Then, once decisions have been made, the model is used as a guide to produce drawings needed for production, future design changes, final surfacing and reference. This "reverse" approach, used by few designers and seldom taught to design students, has its own special values:

- Some designers and some design situations lend themselves perfectly to this method.
- Taught to students, it can reinforce an appreciation of the relationship between two-dimensional (drawing) and three-dimensional (model or prototype) work.
- It is a lot of fun and the exercise is seldom forgotten, even by those who have tried it only once.

GOING FORWARD

In this book, the chapters on constructing finished models (in foam core, foams and styling clay, for example) all indicate the classical approach in design: building a model to conform to specifications in a scale drawing. The work usually begins by deciding on a scale for the model, producing a drawing at that scale in several views, then dividing the drawn model into sections, as seen from one or more views. There must be enough views (front, side, top, etc.) to clearly define the shape of the model. For a symmetrical object, such as a round bottle, two views are enough. Before locating the sections the top view is divided into a series of grid squares. This grid will correspond to a grid marked on a base board. The model will be fixed to this board during its entire construction.

REFERENCE KEEPING

A lot of templates may not be needed for some models, especially fairly simple shapes that are small enough to be hand-held or machine-shaped instead of being attached to a base for building. In this case, instead of sectioning off the drawing, the drawing is dimensioned at any scale drawn. These dimensions are used to locate details on the model, such as holes, cut lines, and simple (radius) curves.

To make such location possible, it is necessary to keep at least one reference on the model being shaped, until all the details have been located and produced. Ideally, the reference is a flat surface (which may be marked for later curving). Sometimes just a straight edge will be enough. With this same surface or edge on the drawing used as a reference, all the details are dimensioned on the model and produced, before eliminating the surface or edge by final shaping. Without following such references (if the model's exterior is entirely shaped to curves too soon), it may be nearly impossible to precisely locate details on a hand-held or machine-shaped model. A flat reference surface also serves as a solid base, without which it would often be very difficult to machine a model.

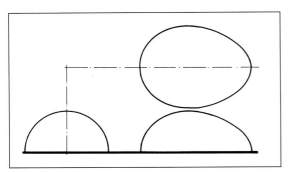

Fig. 2-1. From clockwise, top, side and end views of the egg.

TEMPLATES

Let us assume that the egg has been re-designed. (The following shaping techniques may, of course, be used on more complicated models. The chapters in this book on foams and styling clay also describe template work which may complement the following information.) A three-view drawing of the egg could be similar to Figure 2-1.

Only half the egg is shown, since it is symmetrical. A full model can be made in two halves. Although three views are not needed to define this shape, they have been drawn to give more choice when proceeding with template work.

Once a scale and dimensions are chosen, the sections shown in Figure 2-1 may be used to make templates to guide the shaping work. There are many ways of using templates, which include the following:

- Templates may be male or female, depending on the material to be shaped, ease of work, etc.
- Templates may be permanently incorporated in and become part of the dimensions of the model.
- Templates may be used to occasionally check shaping work.
- Templates may be used as shaping tools for custom work.
- Either hard or soft materials may be used to make templates, depending on their use.
- Reference lines, surface grids, etc. are needed to guide template use.

ROUGH PREPARATIONS

Still building a model of the egg, the model material is first roughly shaped. Using hard or non-malleable materials, such as foam or wood, preliminary cuts are made as shown in Figure 2-2.

Fig. 2-2. Preliminary cuts.

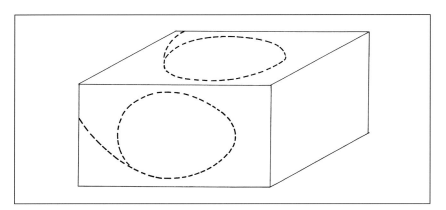

For the cuts, two views are drawn on a squared-off block of material. The first cut takes advantage of the flat bottom of the block, needed for safe and accurate work on the band saw. The second cut is made after replacing (and spot-gluing) the cut-off piece, to provide the flat bottom needed for the cut. This is an old technique—used for cabriole furniture legs, for example—that is still very useful today.

With soft or malleable model materials, such as styling clay or built-up plaster, the work usually begins with an armature, or "buck," over which is laid a thick layer of the material to be shaped back down later using templates as guides or as tools (although some shapes cannot be worked easily with templates). The construction of such armatures is described in this book's chapters on styling clay and foams.

MALE TEMPLATES

One would use a male template when a malleable material is to be shaped into half an egg. A full-sized, fairly rigid (e.g., styrene sheet) template of the egg's base is made from the top view of the drawing. Carbon paper, cutting out and gluing a copy of the drawing, or pricking through along the line with a pin may all be used to transfer the line to

Fig. 2-3. Oversized slabs of foam sandwiching the templates.

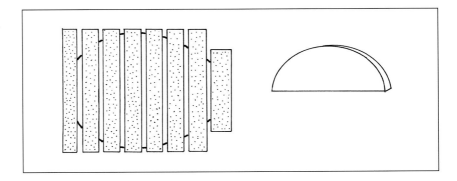

the template to be cut and sanded to shape. The template may be thin, but in any case its thickness will be part of the model's dimensions (its height). This template is glued to a baseboard, and the clay laid over it. Shaping (using female templates) will be guided all around the egg's base by this permanent template.

The sections shown in Figure 2-1 can be used to trace and cut male templates, as seen from the end view(s). The number and spacing of these along a gridded baseboard will depend on the model's curvature. Going further than the egg, frequent curve changes or smaller curves would require more and closer-spaced templates for tighter control of the shaping. The eye and feel of experience can help reduce these. Going back to the egg, if it is to be shaped in styrene or urethane foam, such templates may be made in soft, easily-cut materials such as foam core or cardboard. They will be incorporated in the model (becoming part of its dimensions) and will serve as limiting guides when removing foam. Oversized slabs of the correct thickness of foam are glued together, sandwiching the templates, as shown in Figure 2-3.

When removing foam, the shaping stops at the templates and their surfaces are connected by a smooth contour of the entire egg surface. The surface, keeping the templates in the sandwich, is later filled, sanded and prepared for painting.

FEMALE TEMPLATES

The most frequent use of female templates is with soft, malleable model materials, such as clay. Again with the egg, the end-view section templates described above could be female instead of male. These, in a hard material, such as styrene, acrylic or dense wood fiber sheet, may be used to press into the clay at the correct positions along a grid board. The clay is then shaped down to the indentations made by the templates. This process of fitting (pressing), adding and removing clay may have to be repeated during the shaping work.

Female templates may also be used as tools for shaping clay. Leaving the egg, an example would be a template for a specially contoured section that continues along an axis, as shown in Figure 2-4. Such a template should have big feet, especially if it is to rest on only one foot: A large flat base is needed for precise referencing to a baseboard, and a small template base might too easily rock out of position.

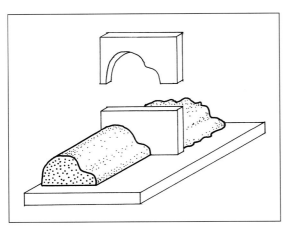

Fig. 2-4. A female scraping template shaping clay.

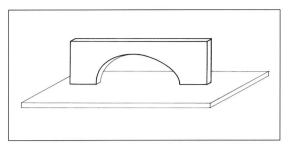

Fig. 2-5. A half female egg template.

Back to the egg, another example is a female template of the side view (or half the top view). This template would look like Figure 2-5. Both pressing and scraping work could be done with such a template, rotating it around the centerline along the egg.

No matter what type of templates are chosen for the work, they should be traced, cut and sanded carefully. Jigsaws and band saws may be used to cut the sheet, inside or outside the traced line. Then the template should be sanded smooth down to the line. When the model is to be filled, primed and painted, the templates need not be sanded to the model's final finish. In fact, they should be one or two millimeters smaller than the final model's surface, depending on the thickness of all the surface treatments that will be coating it.

These descriptions of a few simple template types represent typical approaches. Much more complex shapes than the egg may be made using male or female templates. A little planning and imagination with each new project will reveal ways to make shaping work easier and more precise. Again, there are many possibilities, so many that even a complete book on template use would not cover all the new design model ideas that may come up.

ADJUSTABLE TEMPLATES

The two main types of adjustable templates used for shaping or drawing are:
- The "snake," a semi-rigid, semi-flexible ruler that can be bent around a shape. It has no "spring-back," so it copies the shape exactly.
- The contour gage, a row of pins along a frame in which they can slide up or down. The rack can be pressed to fit a contour, then traced or held against a model (for example, a wood turning) to check the shaping progress.

Both types of adjustable templates may be used to go from the drawing to the model, or vice versa.

READY-MADE TEMPLATES

For flats, angles and circles, there is no end to what can be found for sale ready-made or already at hand. Such templates are also quick to make using the most basic measuring tools.

Then again, it's not every designer who wants to limit his or her design vocabulary to radius curves and flats, no matter what a teacher has shown or a production engineer is willing to allow. There is still a whole world of fantastic shapes to be created or discovered and special templates can be made for all of them. Some very specialized templates may even be made by plotting curves of mathematical functions, where freehand drawing or available sweeps would limit the work. One such example was a three-meter parabolic mirror for a solar-heated furnace, made of mirror styrene sheet bent into and attached to female parabolic frames. The prototype was used to produce instant hot water.

GETTING THERE IN REVERSE

There is the other—reverse—approach to design in three dimensions, as the following section describes. Designing directly with a model is not only valid, but is also an experience well worth trying.

THE ELEMENTS OF FORM

It is surprising how few basic design elements there are in three-dimensional forms. It is also satisfying to see that thousands of years of art and design have not depleted the possible combinations and developments of these few elements in shaped works. These simple elements are:

- The edge, or corner, which can be either exterior or interior (as a groove or a cut line). The edge may also be sharp or dull, and this sometimes depends on the angle made by the surfaces that meet to form this edge.
- The flat plane.
- The curve, which may be simple (in two dimensions) or compound (curving in three dimensions), and may be either concave (hollow) or convex (bulging) in its nature. There are no limits, not even mathematical, to the possibilities for curved surfaces.
- The radius curve, mentioned here only because industrial engineers and a few designers tend to rely on it quite heavily. It is a simple curve to draw, economical to achieve in products and limited by its constant: the radius.

DEVELOPING THE ELEMENTS

Artists and craftsmen have always played with these elements, but usually in an unconscious way. They developed these elements from one to another and combined both the elements and their developments into final forms: sculpture, models or functional products. The following study will emphasize a conscious effort involving this process. This may help create an awareness and a control in future work. It should also help prepare for important aspects of today's design: the critique, initial and final presentations, and eventual design change decisions.

For this study, the reader should prepare 10 or 20 small blocks of clay or urethane foam to make a practice run.

Shape each of the elements first (the blocks will already have the flats and some kinds of edges and corners). Try some of the possibilities—sharp or soft convexes and concaves, for example—in different curvatures (ellipse, parabola, hyperbola—there is really no end to these).

Then take blocks and try developments of each of the elements, back and forth, from one to another. List, then try all the possible combinations in varying ways. A few examples:

- An edge, fading slowly into a flat, or a convex.
- A flat, slowly developing into a concave.
- An edge, breaking fast into a flat, a convex or a concave.
- A convex in one direction, concave in the other direction.
- A convex which quickly becomes a concave, both in both directions.

- A convex or a concave that changes its curvature as it goes along.
- A flat that simply twists as it continues.

There are many more possibilities. Try them, keeping all the pieces you shape. If enough developments are tried, you will probably discover some that you like more than others. Keep these in mind as you continue this exercise.

COMBINING THE ELEMENTS AND DEVELOPMENTS

Now take a few bigger blocks of urethane foam, cut in the proportions with which you would like to start. This could serve as the beginning of a model design, but don't have a finished idea in mind. You may end up with a shoe or a potato, a car or a human figure, but for now resist the temptation of even giving a name to your work. Just make the practice run and see what you end up with. Start by combining those developed elements that you preferred—some or all of them—on the same block. See how you like the combination, and start other blocks, to vary the experience. Also, see how other people react to your results.

CRITIQUES

Critiques, presentations and design changes all involve notions of pride, success, economics and the gift of gab. As one designer recently put it, "We are in the business of art."

Solicit critiques and be open to suggestions, being careful not to interpret them as personal attacks. When giving critiques of your own, remember to critique the work and not the artist. Also, when formulating a critique the criteria you consider will help you assemble a kind of design vocabulary.

Do these combined elements and the chosen developments complement each other as they form an entire surface? Do they harmonize or do they

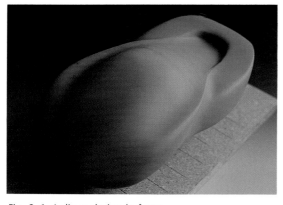

clash? Do they suggest a flow and give an overall connotation of movement? A continuity? Or do the combinations or the individual developments show interruptions? Are there areas of indecision or compromise? Or is an abrupt change desired to give tension to an otherwise overly-soft or boring surface? Keep in mind, too, that critiques are highly subjective; one expert's beautiful curve can often be another's monstrosity.

Figure 2-6 shows a foam model that resulted from experimenting with the elements of form and their developments. As it happens, it ended up looking like a car. Having survived the critique stage, it is now ready for finishing work.

Fig. 2-6. A direct design in foam.

SLICING THE BREAD

Before you rests the sanded foam model of your dreams. It may not resemble a loaf of bread, but you must still take it to the band saw and slice it up as though it were. Start by marking reference lines, which will be needed later. One line, the "waterline," is scored horizontally all around the model, at an arbitrary height that will allow it to cross all or most of the slices. A marking gauge can be used for this, or another

Fig. 2-7. A model in slices.

means is to keep a pencil or a point at a uniform height along a smooth table. Then score a "centerline" along the approximate center of the model, on the bottom or the top. Lightly glue the model to a squared-off piece of foam core or cardboard, and glue foam wedges under any open spaces to provide a firm, flat support for cutting each slice on the band saw. The board should extend far enough out from the end of the model to be used as a safe handle for cutting the last few slices.

Set up the rip fence on the band saw to cut even slices about $3/4$" (2 cm.) thick for a model about $11 3/4$" (30 cm.) long. If there are a lot of details or tight curves in parts of the model, thinner slices will help achieve better definition of those areas later. Number the slices, either before or as they are cut off. One former student forgot to do this, and his finished model—after a lot of shuffling of slices—looked quite different from his original design. Slice up the model, then finish the work by cutting a piece of foam into slices as thick as the groove that the blade makes in the material. The rip fence is set up to cut these to fit into a test cut by the blade used (every blade will cut differently). You will need these thin slices later if you want the final model to be as long as the one you started with. Figure 2-7 shows the model as it would look sliced.

THE THREE-VIEW DRAWING

Using always the same side of each numbered slice as a template, numbered outlines of all the slices are traced on paper. With tracing paper, carbon paper or by pin-pricking, these outlines are superimposed on one sheet of good-sized drafting paper. Prepare the paper beforehand: lay out a space for the end, side and top views of the model (the end and side views should line up). Two end views can be made for model shapes whose end-view outlines merge too much and are confusing when superimposed. Colored pencils may also help keep separate sections clear. Draw a centerline through the top and end views, and a waterline through the side and end views, to line up the traced outlines of the slices. Prepare a space for the drawing's title block and another for eventual explanatory notes and design changes, just as you would if you were manufacturing this product for years to come.

The superimposed outlines, numbered from the first to the last slice, make up the end-view sections of the drawing. Broken lines are drawn to represent hidden outlines. Then go to the side and top views and draw vertical lines to represent the slices and the cuts, distanced at their correct thicknesses.

Project points from the end view to the side view. For example, points at the top and outside of each slice outline, and points at sharp breaks in line direction will help define the side view. Each point on each end-view section is projected horizontally to its corresponding slice line on the side view.

Do the same for the top view. This view can either be projected in 90-degree turns from the end view, or another end view can be drawn, at 90 degrees from the first, then projected straight to the top view. The same points of the end-view sections will define the top view.

Connect the projected points on both side and top views by drawing smooth, continuous curves through them, ignoring any points that fall outside the line. A French curve or sweeps will help to draw the lines. Irregular points might show poor model work, but they are most often the result of inaccuracy in slicing, tracing and projecting, so such irregularities are best ignored. Figure 2-8 is a three-view drawing of our model.

Fig. 2-8. A drawing from a model.

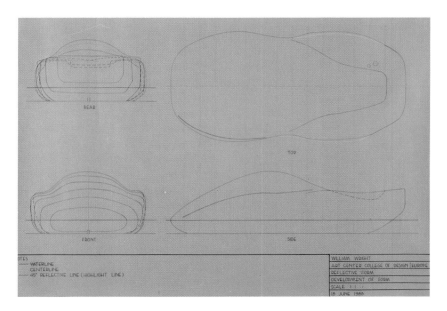

The drawing, with its vertical (and horizontal, if desired) section lines may be used for further study. Design changes decided during a presentation may go through the drawing, be documented as revisions and end up translated on the model. Local, piece-by-piece surface corrections may also be made through the drawing to the model, using geometrical techniques (triangulation and parallel key projection) for proportional curve development, or with modern means, such as computerized calculations of curve tendencies, by differentiating the mathematical functions of portions of the curves.

THE SKELETON, OR SPACE FORM

The finished drawing may also be used to construct a space form of the model. This, somewhat like a skeleton or a shell of the model, is used in some industries for further design studies. In the automobile industry, important work takes place inside such spaces. The comfort of "average-sized" passengers and the functionality of controls are objects of intense study and frequent design changes. In the manufacture of electric and electronic equipment, studies of the interior configuration of parts are conducted for volumes needed, ease of service, heat buildup, etc..

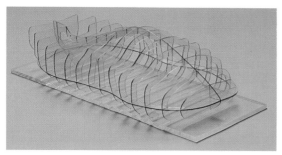

Fig. 2-9. A space form in a material that shows space.

There are many materials and techniques which may be used to build space forms. Figure 2-9 shows a painstaking construction in transparent acrylic sheet, built from the drawing described above. Other materials may be used for the space form, depending on the size and the use intended for the construction. Foam core, styrene sheet, wire and balsa wood have all been used successfully. The construction can resemble the kits children buy for model airplanes, except that nothing is pre-cut to fit. Measuring, cutting, fitting and gluing have to be precise, and in planning the structure extra strength is needed for certain uses and studies.

FINISHING THE MODEL

The slices of the model may now be put back together and the model finished. The slices should still have their numbers and reference marks for correct alignment. Each slice is glued to one of the thin foam slices before gluing to the next slice. It doesn't matter if these thin slices stick out from the model's surface—they can easily be sanded down. What is important is to keep the glue inside, away from the surface, since glue sands less easily than foam and causes ridges to develop on the surface. Once together again, the model is coated with filler, sanded and maybe filled again and sanded again, primed and painted, following the techniques described in the chapters of this book on foam models and on finishing and painting. Figure 2-10 shows the final result of the model work.

If the model is gloss painted and polished, a "highlight" study can be made, preferably outdoors. The lines or streaks of reflected light will reveal not only the smoothness and continuity of the surface, but also some useful ideas on how the surface treatment—the elements of form and their developments and ensemble—can affect the appearance of the model in good light. This study may be a surprise; physical surface intentions are often very different from the play of light they create (for example, a straight horizontal ridge or edge along a model will often give a sagging highlight line). Right to the end of the study, there can be more reasons for design changes and opportunities for planning the next project.

Fig. 2-10. A finished foam model.

CHAPTER 3

CHOICE OF MATERIALS AND TECHNIQUES

Photo, design and model: Ulrich Schraudolf.

MATERIALS

Although there are many materials used for design modelmaking today, there is surprisingly little to consider when choosing from them. Deciding on techniques, however, may be somewhat more difficult.

The easiest way to choose materials for a design model is to do so with the model's future purpose in mind. In other words, does it need to be an extremely finished model, or can it look more like a work-in-progress? Who will be seeing it, what will they be determining, will it be shipped and, if so, how often? What may complicate the choice are the preferences of the modelmaker, the designer or a manager. One may want to try a new material, another may already have a favorite and another may lack modelmaking knowledge. The most important considerations are cost of materials, surface(s) of the model, finish, details, health risks, time available, form development, shipping and storage, presentation and design concept status. The glossary at the end of this book lists most of the materials used for modelmaking today, along with indications for their use.

COST OF MATERIALS

The cost may be very important for a client's or design student's budget. Some materials, such as silicone rubber, dense urethane foam and cast urethane, cost so much that it is often worth trying to substitute a cheaper material, even if it takes a little more time to get the same result. Some projects may be so limited that only a quick mock-up in a cheap, fast material should be considered.

Fig. 3-1. Telephone in cast urethane, acrylic and acetate. Photo, design and model: Philippe Holland.

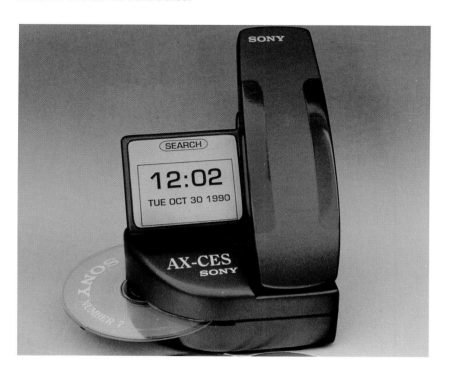

Fig. 3-2. Foam telephone. Photo, design and model: Robert Jurgens.

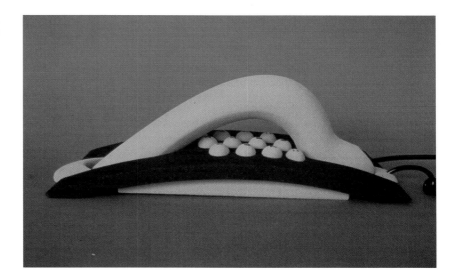

Fig. 3-3. Corian® speakers. Photo, design and model: James Hamilton.

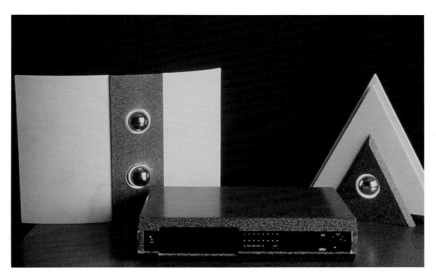

SURFACE

For a model with mostly flat, smooth surfaces, you should select a sheet material rather than shaping and finishing another material to get the same results. For compound-curved surfaces, materials such as foam core simply will not work.

FINISH

When a matt finish is desired, it is quicker and cheaper to use a porous, naturally-matt material like urethane foam or PVC foam sheet, rather than a hard material which must be rendered matt. (See Chapter 6 for information on the special treatment of matt urethane foam models.) The same may be true for brilliant finishes or other special effects sought with a surface finish. Very few metallic paints look like real metal, so it may often be better to use the real thing for those models. Transparence, too, is hard to simulate with even the best rendering or other graphic applications. Sometimes clear acrylic (or styrene or PVC) is the only way to go. For many gloss surfaces, choosing between softer materials (e.g., urethane foam) and harder ones (e.g., solid cast urethane) is not always easy. When making a small model, it often takes about the same amount of time to completely finish it, whether it is made of soft or hard material. Foam can be shaped and sanded quickly, but requires surface filling, more sanding, and often even more of the same, before it can be primed and painted. This can take a lot of time. While a hard material does take longer to shape and sand, once this is done it is ready for finish coats, with almost no more surface filling or sanding.

DETAILS

Harder materials allow for finer details. There is no use trying to produce fine details and cut lines in soft foam, for example. Not only is the material too fragile for fine shaping, but it is also porous, requiring heavy coats of filler and primer to fill the pores. These coats will just bridge over any of the details created, and require difficult repetition of them on the finished surface. The better option is to use a hard, non-porous material for detail work, then finish it with very thin coats of paint that will not hide all the work done. For cut lines, the best approach is often to separately shape and finish parts, which are then joined after painting. This approach not only avoids problems, but produces real cut lines. It also avoids masking lines from the multicolor painting of one piece. Masking lines are always visible, no matter how much care is taken to hide them, especially on small models that must be presented up close. Separate parts work is just as easy, often easier than, creating cut lines. It may also be a big step toward making very professional models.

Fig. 3-4. Men's toilet accessories in hard polyurethane. Photo, design and models: Tevfik Fikret Uçar.

Fig. 3-5. Flower shop stool prototypes in fruitwoods.

HEALTH RISKS

The health issues concerning modelmaking are becoming more publicly recognized every day. Manufacturers have not always done a good job of informing us of the ecological consequences of the use and disposal of most materials, and we already know that we can't simply rely on suppliers' labels and brochures. It may be time for everyone concerned to get better, truly impartial information, then start making choices that go beyond the typical "you need this" advertising. This may involve being open to some very simple chemistry, physics and nuclear science. An effort has been made throughout this book to increase a sense of responsibility and awareness in working with model materials. But this is only the beginning. There is still a long way to go, and everyone must take part.

TIME AVAILABLE

There is often little difference timewise in working with hard or soft materials because of the surface treatments needed with soft and porous materials to achieve some desired finishes. Sometimes it is very clear: time or budget will only allow a fast mock-up in a material such as foam core for a study. At other times, there may be a decision to make. For example, with only one day left to finish, if the model is still damp from wet sanding, it may be a very good idea to present it as is, rather than risk a paint job that may bubble and blister during its presentation.

FORM DEVELOPMENT

Soft materials are easier to sculpt than hard ones. Paper materials do not allow for compound curves (foam core is no exception). Sheet metal requires special forming equipment as does thermoplastic sheet, which requires a forming machine in most cases, and usually a master mold. There are also design limitations in forming metals and in thermoforming. Styling clay is a great favorite of car designers, but it is slow to work. For design changes, especially add-ons during modelmaking, styling clay is probably easier to use than foam. Plaster is still used in some studios, even for big work. As messy as plaster can be, it is much cheaper and less of a hazard than urethane foam, and smells a lot better than styling clay. Curved forms may be produced with equal ease on materials such as clay, foam and plaster.

SHIPPING AND STORAGE

It often happens, after the "final" presentation, that a manager or client has further plans for a model. This may mean long-distance shipping and storage under risky conditions. Materials such as styling clay and foams are not really suited to such exposure. For many modelmaking materials, it may be necessary to build a hard reproduction of the model, in fiberglassed resin, for example, for such use.

PRESENTATIONS

There are some considerations in planning a model which depend only on the presentation(s) expected. Is it just a quick, cheap mock-up needed for a better idea of an ongoing study? For another project, maybe well-taken photographs are enough. If the model is big, like a fifth-scale automobile or a tenth-scale railway car, fine details may be unnecessary. If included, they might even hurt the model's presentation with their "child's toy model" appearance. Viewing distances and light can be controlled for big model presentations,

Fig. 3-6 and 3-7. Egg packing. Photo, design and model: Jerome Eichmann, Double Crème Communications, Switzerland.

Fig. 3-8. Urethane telephone. Photo, design and model: Caroline Sechehaye.

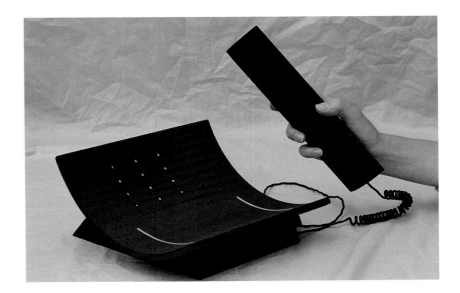

Figs. 3-9 and 3-10. High density polyurethane foam can opener. Photo, design and model: William Wright.

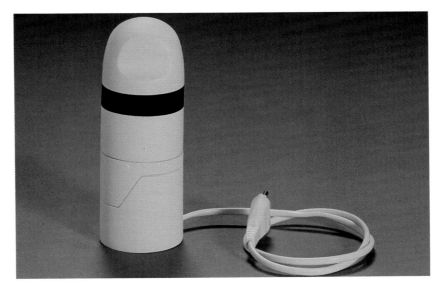

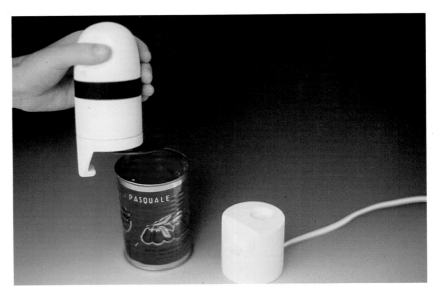

sufficient distance alone being sought for a better appreciation of proportions and surroundings. It also helps keep viewers' hands off the model. If there are moving parts on the proposed design, such as open and closed positions of some parts, it may be enough to build the model in several parts (without the eventual mechanism) and simply present or photograph it in each position. There is also the question of fragility. If the presentation is to be from a distance or close-up, there is always the chance that someone will want to touch it, and not all materials react in the same way.

DESIGN CHANGES

Are the designers really finished designing the design? And the clients? Does everyone really want to start all over again with a new model? If design changes are expected on the model, a material such as unpainted styling clay or urethane foam (not surface-filled) will allow for the easy addition or subtraction of design changes.

OTHER CONSIDERATIONS

Transparent parts on the finished model usually call for clear plastics, though economy may limit the choice. Found or ready-made parts may be just the thing for certain parts, instead of making them. For a given model, or parts of one, it may be necessary to choose between a thermoplastic and a thermoset. For example, a sheet that must be thermoformed into a compound curve may be one of many types of plastic, but not cast polyester resin sheet. Or, a plastic material used for a part that must be exposed to heat or solvents during or after the model work should either be a thermoset or at least be able to resist the temperatures and solvents anticipated.

Fig. 3-11. Bathroom project. Photo, design and models: Ulrich Schraudolf.

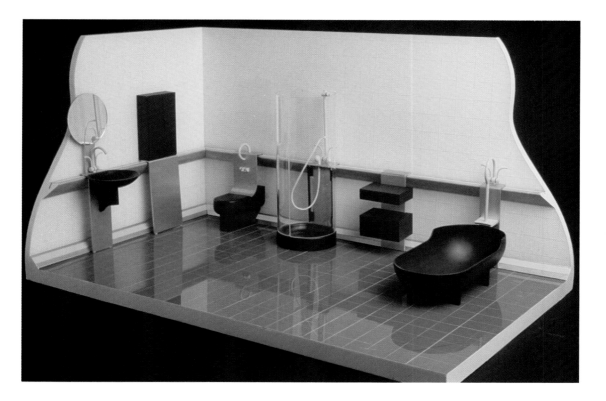

Fig. 3-12. Perfume bottle in mixed materials. Photo, design and model: Vibeke Nodskov.

More specific examples could be given, but mostly they are variations of all the above considerations. Skill will come with practice and with the knowledge that small failures are just as important in the learning process as big successes.

READY-MADE OBJECTS AND MATERIALS

One former student was running out of time with her telephone model, so she decided to try the patience of a local pharmacy. She insisted that she wasn't at all sick, but simply needed 8 mm. pills. They turned out to be perfect for the telephone's buttons. Wooden dowels, steel cable, iron capnuts and other assorted hardware and fittings, real rubber or leather, patinated lead and copper sheet, aluminum, selected hardwoods, barbecue motors, PVC pipe and plumbing fixtures, plastic fishbowls—all these have been used very successfully as parts in models. The possibilities are truly endless. Sometimes using the real thing is the only way to go, rather than "reinventing the wheel" or attempting to imitate with rendering, painting or graphics. A found object or material might be exactly the way to put originality and imagination into a model design, and it may also save some time in the process.

MIXED MATERIALS

Very often, studying how to proceed with a model will show that several materials are to be used. For example, a major part of the structure may be shaped in urethane foam or plaster, then filled and painted. A part with fine cuts or details may be made separately in cast urethane. Transparent or thin-walled parts may be vacuum formed in acrylic or styrene. Tiny parts, such as push buttons, may be ready-made using pills or bits of wooden or acrylic rod. Graphics may be produced on thin acetate sheet, using the many processes available, or creating new ones for specific jobs.

Fig. 3-13. Perfume and toilet accessories in mixed materials. Photo, design and models: Vibeke Nodskov.

Fig. 3-14. Acrylic kitchen. Photo, design and model: Caroline Sechehaye.

TECHNIQUE CRITERIA

A designer, student or client will often come back from seeing a project with a new idea even crazier than the last one. The questions concerning the making of the model may be, "Where to begin?" or, "Which way to go?" Many of the answers are tied to the choice of materials. Some of the criteria are the same. There are, however, other things to consider when deciding on techniques.

MODELMAKING EXPERIENCE

Are you capable of building a good model using an unfamiliar technique, or should you stick to one that you know well? Do you prefer or enjoy one technique more than others? One former design student fell in love with turning on the lathe and proceeded to make her entire model of an orange juice squeezer with turned parts. The choice is really up to you.

SHOP CAPABILITIES

Is the model shop well enough equipped to do the job the way you've chosen? There is no use hoping it is, or starting a project with the assumption that the equipment you need will be there. It may be necessary to go out and look elsewhere when doing special jobs. The investment required for outfitting a good model shop is an ongoing concern. Often it boils down to what type of work is done most often, since there is usually a limit to the funds available for new equipment. Space is also a common problem, as are the restrictions of organizing it for safe working conditions. A sandblaster may be just what you need for a project, or a sheet metal bender or a high-volume paint spray system for painting walls. But this material, once bought, may sit around for years without being used.

END RESULTS DESIRED

The choice is obvious, in the case of transparent or identical parts, where vacuum forming or casting and molding may be the only choices. In other cases, there may be some apparently equal decisions to make. There may be no fast answer, and several ways to proceed. In such cases, try the fastest method first. If it doesn't work, at least you will find out quickly that the time was wasted. This can often happen, even for individual operations that go into making a whole model. Try and fail if necessary, but do so efficiently.

TESTING AND EXPERIMENTING

Since design modelmaking is always full of surprises with new materials and new concepts, be prepared to test and experiment. Always test an unfamiliar material, and always try out a new operation on scrap material or on a hidden part of the model—even one as simple as drilling an angled hole—rather than risk ruining it. As with choosing materials, deciding on techniques will become a question of practice and experience.

SPECIFIC TECHNIQUES

Parts of other chapters in this book describe techniques used with various materials. The following is an outline of specific techniques to employ for various situations.

SHAPING

Shaping is most often hand work for models or parts of complex or original forms. This usually excludes hard materials, such as concrete and metals. Handshaping is not often attempted for flat or simple, symmetric surfaces that are readily available or more easily produced by machines (turning, milling, planing, etc.). Always be sure, whether shaping by hand or by machine, to plan and keep a reference surface, or at least a line, on the model until all holes, dimensionally-spaced details and the like are finished.

CASTING AND MOLDING

This work is usually reserved for multiple, identical parts where repeated originals would take too much time or lack precision. It is also used in special cases, such as for transparent parts, where shaping a clear model would be very difficult, or when a resistant model in a hard-to-shape material is needed.

THERMOFORMING

Used only with thermoplastic sheet, thermoforming is used for multiple, identical hollow parts, and for smooth, compound-curved surfaces too difficult to shape or cast. The design limitations and the many "tricks" should be kept in mind for successful forming. Be prepared to experiment and waste a few sheets of material for each new idea.

SEPARATE PART CONSTRUCTION

Where cut lines are shown on models—representing assembled parts of finished products—build the model by shaping and finishing separate parts, paint them separately and then join them together. This eliminates masking lines for several colors. Also, details such as fine radius (or other) curves approaching the cut lines are saved from fillers, primers and paint, which tend to bridge over and hide such work. The model itself is often more easily made this way than by planning and shaping one piece.

FINISHING AND PAINTING

Sometimes it makes little sense to shape, finish and try to imitate a surface when such a surface is already available in other materials. Flat, gloss, matt, cloth and rubber surfaces are examples of when time can be saved, and better effects produced, by using the "real thing" or a material already manufactured.

ARCHITECTURE MAQUETTES

The model work usually done for architectural presentations is somewhat specialized in design modelmaking. The typical approaches, as well as the results, are surprisingly limited. In this case the model work is characterized by one extreme or another: abstract, primitive work, even for important public projects, often done only in cardboard; or meticulous models in one material, such as acrylic sheet, for simple "rabbit cage" apartment buildings.

This tendency can probably be explained by the fact that in both the United States and Europe, most architecture (and interior architecture) schools place little or no emphasis on presentation maquettes and offer no instruction in the work. Good maquettes are sometimes made, however, as the following photographs show. These may be even more important in architecture than in other areas of industrial design, since the final product will be universally visible for a long time. And here the same rules apply: a better appreciation of the proposed work is had with a model than with drawings, computer renditions or written specifications.

Fig. 3-15. Architectural maquette by Laurent Zali.

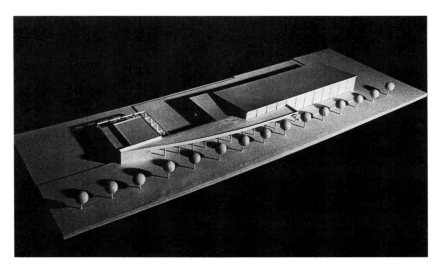

Fig. 3-16. Architectural maquette by Laurent Zali.

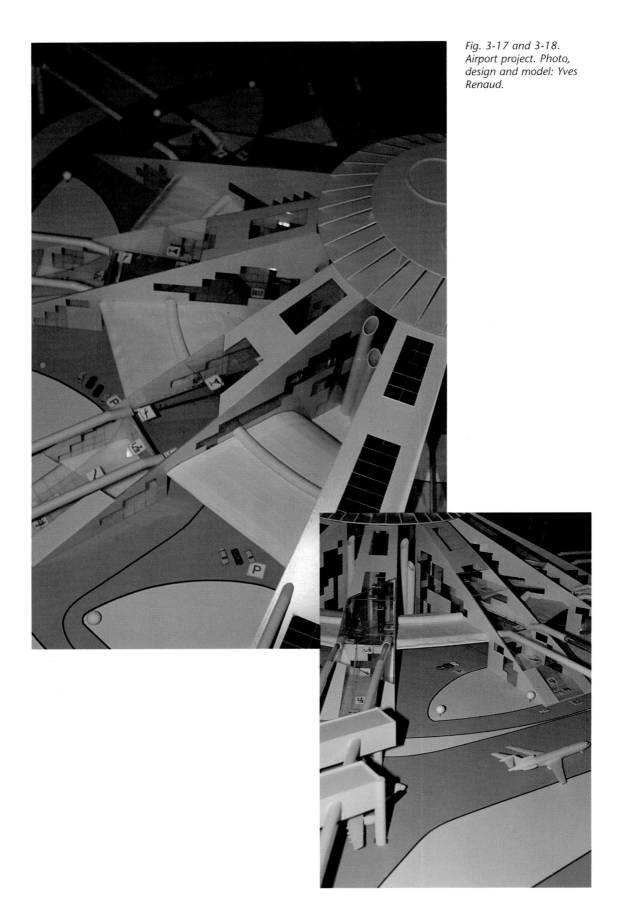

Fig. 3-17 and 3-18. Airport project. Photo, design and model: Yves Renaud.

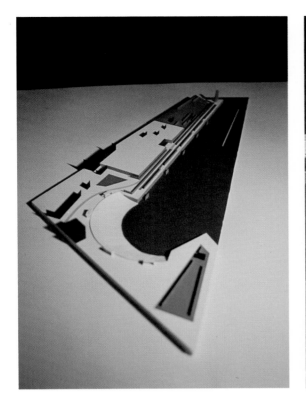

Figs. 3-19, 3-20 and 3-21. Architectural maquettes by Patrik Vuillemin.

CHAPTER 4

PHYSICAL AND CHEMICAL PROPERTIES OF MATERIALS

PHYSICAL PROPERTIES

Not all materials were created equal, so knowing about their simple properties may come in handy for the designer or modelmaker. No attempt will be made here to present advanced physics or chemistry. At the very least, you can learn a vocabulary of basic materials technology so that production engineers won't wipe the floor with your design proposals. More seriously, a knowledge of material properties will very simply allow you to know what can or cannot be done when building models or designing products. This chapter presents the main physical and chemical properties of materials, with discussions of their advantages and disadvantages. Note, also, that no attempt is made to list and completely describe *all* materials. Such information is readily available in manufacturers' brochures as well as in engineering handbooks.

HARDNESS

A material's hardness is its impact and abrasion resistance, and is often related to brittleness, lack of bending or tensile strength. Some very hard materials are diamond, glass, cast iron, polyester resin and other thermoset resins.

DENSITY

The density of a material is the mass ("weight") of a unit volume of the material. It is not very precise to talk about "heavy" or "light" materials. There is a great range of densities, from dense metals like gold (about 30 grams in one cubic centimeter) to low-density materials like polystyrene foam and thixotrope powder.

TENSILE STRENGTH

Steel, spiderweb, carbon and glass fibers all have high tensile strength because they resist a pulling force. Clay, concrete and as-cast resins have low tensile strength.

COMPRESSIVE STRENGTH

Concrete and stone, acrylic and polyester are strong under compression. Rubber, lead and polystyrene foam are not. The typical examples are in building: there are suspension bridges built with thin strands of steel, and tall buildings compressing many layers of heavy concrete. But the opposite is never attempted.

ELASTICITY

Rubber (including silicone) and spring-tempered steel are very elastic. They can be stretched or bent, then returned to their original shape. Stretched beyond their elastic limits, they stay there, deformed. But within the limits, they can be worked over and over until they break down from "fatigue".

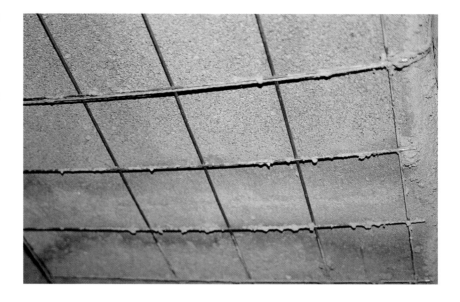

Fig. 4-1. Wire reinforcing for cementing over a chimney. The cement gives the compressive strength and the steel wire the tensile strength needed to avoid bending in strong wind. Photo: Monique Carroll.

DUCTILITY

This is the ability of a material to be pulled into fine strands without breaking. Copper and soft iron are ductile. So are many thermoplastics when softened with heat for thermoforming. Ductility can be an inconvenience as well as an advantage. Some glues, tar and lead (on roofs) will "creep" with constant tension. Many thermoplastics will do the same.

MALLEABILITY

Some materials, such as copper and gold, are very malleable because they can be beaten or pressed down into thin sheets without breaking apart. Softened styling clay is, too. Carbon steel is only malleable (and ductile) while it remains in its annealed state.

THERMAL EXPANSION

Every material expands or contracts a specific amount with changes in temperature. This must be taken into account with all products that require tight-fitting parts. It is especially important for big constructions, where such movement could be destructive.

FLASH POINT

The temperature at which a liquid or vapor can ignite or explode is its flash point. With many modelmaking solvents, this point is very low and great care is needed in working with such materials. Even a spark from a tool or a motor's brushes can cause a serious accident.

MELTING POINT

All pure solids will melt at specific temperatures. Ice, for example, melts at zero degrees celsius. This is also the freezing point of water. Thermoplastics, too, have their own melting and softening temperatures. PVC, for example, can be thermoformed at a much lower temperature than acrylic, making it easier to work with.

BOILING POINT

Liquids boil at specific temperatures. They will also evaporate slowly below their boiling points if they are exposed to air. Water trapped under a skin of paint, for example, will most likely blister the paint to get out. In thermoforming, water may boil in such plastics as acrylics and polycarbonates, causing bubbles that ruin the transparency.

ELECTRICAL CONDUCTIVITY

The electrical conducting, or insulating, property of a material is often important in design. Not only are good conductors (e.g., gold, copper and aluminum) needed in electrical or electronic parts, but good insulators (e.g., glass, plastics and air) are needed to prevent accidents and conserve energy. Carbon found in fishing poles, for instance, could lead to electrocution under the right conditions.

THERMAL CONDUCTIVITY

A material must be a good conductor to spread heat in cooking utensils, water heaters and flat irons. Copper is one example. A good thermal insulator is needed in buildings, frozen food packaging and cold weather clothing.

MAGNETIC PROPERTIES

Door closures and many fixtures make increasing use of magnetism. Only a few metals are attracted to magnets, which may be permanent (hard steel) or temporary (soft iron), and are activated by a temporary electric current.

VISCOSITY

The resistance of a liquid to flowing is its viscosity. This is often important in clear resin casting, where a very low viscosity is needed to allow bubbles to escape, and in quality paint spraying, where the paint viscosity affects how it coats the surface. Thixotrope powder can complicate this property by helpfully giving liquids a "temporary viscosity" for easy use.

OTHER PROPERTIES

There are other physical properties, such as compressibility (of gases), specific heat (heat capacity), taste, color and smell. A basic book on physics can easily help you complete the list. What is important is to realize that design and modelmaking can benefit from a knowledge of material properties in order to meet specific functions.

CHEMICAL PROPERTIES

SOLUBILITY

Solubility often affects work with solvents and other chemicals. Solvent bonding uses the specific solubilities of plastics to make strong joints. Paints, resins and glues also have specific solubilities (for example, water won't wash off sticky urethane glue or polyester resin). See chapter 7 for more information on solubility.

REACTIVITY

A chemically reactive—as opposed to stable—material reacts with other chemicals, making it useful for some functions but also dangerous as some of those "other chemicals" may be inside us. Many products require materials that will resist reacting, such as rustproof sinks and auto parts, building exteriors and underground systems.

REACTION TIME

Chemical reactions often take place within a specific time. Plaster sets in a few minutes while iron rusts and concrete hardens over years. Catalysts speed up slow reactions, such as the polymerization of resins. Too much catalyst, however, promotes a get-away reaction that can ruin resin casts. Reactions can also take place in the sun or under another source of natural or artificial radiation.

TOXICITY

A material's poisonous properties may be related to its reactivity, but not always. Carbon monoxide gas is harmful because it excludes oxygen from lung cells. Mercury and other heavy metals, for example, are not highly reactive, but may remain in the body for a long time and act as "cumulative" poisons.

RADIOACTIVITY

On a biological level radiation can cause the same reactions as some reactive chemicals. Common product uses for radiation are found in smoke alarms, food density detectors and resin curing systems.

SPECIFIC REACTIVITIES

Aluminum reacts with sodium hydroxide, copper with ferric chloride, iron with nitric acid. All of these reactions are used in the chemical etching of metals. Silver reacts with oxygen in light, producing black silver oxide in exposed photo film. Iron and steel rust if not protected from the oxygen in air. The list of specific reactions is long and not for anyone to memorize. Consult a chemical handbook or an engineer when in doubt.

Photo, design and model: William Wright.

CHAPTER 5

FOAM CORE

COMMON USES
AND POSSIBILITIES

For quick mock-ups, graphics work, interior and exterior architectural maquettes, packaging models and displays, foam core is one of the most useful materials available. Even for some (or parts of) finished models in product design, foam core may be the best choice for the time, budget and work facilities involved in the project.

The material is a sandwich of foam between two layers of fine cardboard or heavy paper, available in various thicknesses, usually from 3 mm. to 10 mm.. Its quality may vary depending on the manufacturer, in the type of paper used or the kind of foam inside. This is important to remember, because some papers bend better than others and some foams will dissolve in certain glues and melt with heat.

Since it is a sheet material, foam core is most often used for flat work. Its paper surfaces prevent it from being bent into compound (convex or concave) curves. But with a few simple techniques, much work in bends or simple curves can be done on a foam core model.

Figs. 5-1 and 5-2. Partly foam core model executed for the exhibition design competition of the Barcelona Olympics using the torch theme. The finished work, exhibited in the Olympic Museum in Lausanne, Switzerland, was accepted for the Barcelona Olympic Games. Design, model and photo: Jerome Eichmann. Finished work: Jerome Eichmann and Philippe Borys.

THE BASIC RULES

CLEANLINESS

Avoid food, drink and the typical model shop, where dirty tables, hands and materials are unavoidable. One of foam core's big advantages is that it can be worked with very few tools in a clean office, studio or home.

A GOOD, SHARP CUTTER

The best cutters are those with retractable blades, for depth-guided cuts. Use a cutting surface to protect the table or desk and use only a very sharp blade, either by changing the blade or by breaking off the dull point on some cutters. A dull blade is still sharp enough to injure someone, however, if carelessly discarded.

SANDING EDGES

Sanding paper edges is necessary if a dull blade has made a ripped cut, or if there was inaccurate measuring or positioning for glued joints. Use fine abrasive paper (no rougher than no. 150) and sand the edge at an angle from the paper surface to avoid scratching it.

GLUE

Some foams used in foam core will dissolve in certain glues and will melt with the heat of a hot-glue gun. One such foam is polystyrene, which is often used in foam core. Make a test on scraps of the material before doing any cutting or fitting work to see whether it melts easily and dissolves in acetone or model cement. Almost any glue will work with polyurethane foam core. Hot-melt glue, wood glue and model cement in tubes will all hold glued joints well. Apply just enough glue to hold the structure together, and always try to glue behind the visible surface. Don't use too much glue. This is one of the disadvantages of the hot-glue gun: it is almost impossible to apply just a small drop.

NO PAINT OR COLOR

Painting is usually attempted to hide mistakes—and it doesn't work. If white or black foam core is not acceptable, use another material. Slight pencil marks or smudges can be removed with a nonabrasive eraser. If a surface is scratched or scraped during the work, it is best to replace the part with a clean one, rather than hope the problem can be corrected on the finished model.

TEST BENDS

Some papers bend and fold more cleanly than others. Even with some good quality papers, the fiber direction allows for clean bends and folds in one direction only. Experiment to find this direction, then mark the sheet for cutting.

TECHNIQUES

Fig. 5-3. A depth-guided cut. Photo: Corinne Neuffer.

Fig. 5-4. Scraping away the foam. Photo: Corinne Neuffer.

Combining the following few techniques will allow for a surprising variety of model work. Practice each technique on scraps to get a feel for the material before starting the model.

THE OVERLAP JOINT

A clean edge or corner can be made using an overlap joint. Only the smooth paper surfaces will be visible along a joined edge (or even a free edge), hiding the foam inside. Plan the work, including the thickness of the foam core in the dimensions. Lay out and mark the sheet for cutting with very fine lines, using a hard 4H pencil (a soft pencil mark may be as much as 2 mm. wide, and on which side of such a line would you cut?). One joining part will be cut as in Figure 5-3, using a retractable cutter set to a depth to cut only through one paper layer and just the foam inside, leaving the other paper surface uncut.

Scrape away the foam between the cut and the edge. When the two parts are joined and glued, a continuous surface will result, requiring only a little sanding of the edge.

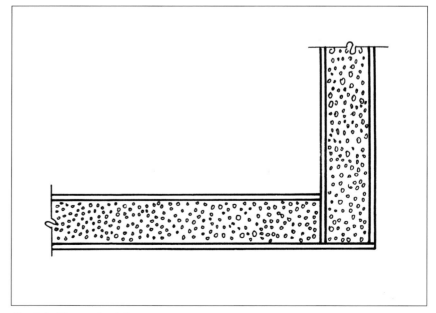

Fig. 5-5. The overlap joint.

THE MITER JOINT

The 45 degree, or miter, joint also makes for a clean joined edge, but it is harder to cut accurately than the overlap. When attempting this joint and its required cuts, use a 45 degree guide for cutting. Otherwise, keep the joints (and their cuts) short.

Fig. 5-6. The miter joint.

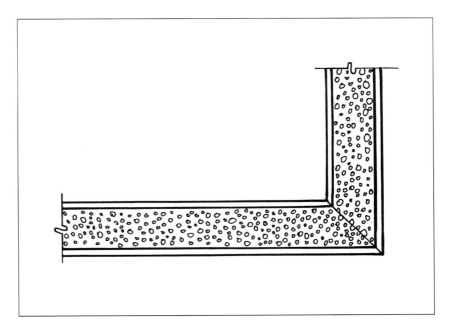

THE BUTT JOINT

The butt joint is the easiest to cut with any material. This joint is used only for interior parts with foam core, such as glue braces, where the foam will not be visible.

Fig. 5-7. The butt joint.

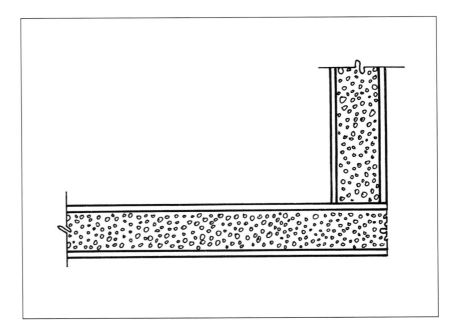

THE SHARP BEND, OR FOLD

A fold in foam core may be used instead of the clean joints described above. It makes a super-clean, tiny radius corner, at 90 degrees or any angle where a smooth, continuous paper surface is desired. Mark the line on the inside of the fold. Then run a blunt tool (the cutter handle is good) along this line to press the foam down to the outside paper surface. This will look as in Figure 5-8.

Fold the foam core along this line. Use glue on the inside of the fold to keep its shape. In some cases, it also helps to glue several foam core braces inside, spaced along the fold. The final result is shown in Figure 5-9.

Fig. 5-8. Preparing the fold. Photo: Corinne Neuffer.

Fig. 5-9. The fold. Photo: Corinne Neuffer.

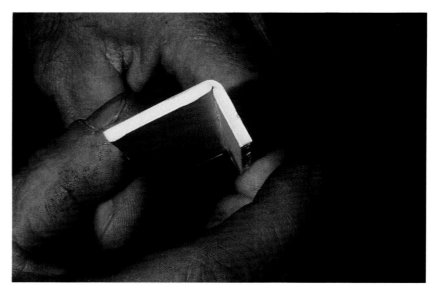

THE SOFT BEND, OR SIMPLE CURVE

Any simple curve or bend can be made with foam core. A simple curve here means a bend in two dimensions. A compound curve—in three dimensions—is impossible to acheive with foam core. Other materials and techniques, such as shaping foam or clay, or thermoforming sheet plastics, are used for model parts requiring compound curves.

Prepare the bend by marking the cut lines at each extremity. Then cut through only the inside paper layer along these lines, and peel off the paper between these cuts, as shown in Figure 5-10.

Be patient when peeling away the paper because it doesn't always want to come off in one piece. Fingernails are as good as any tool for this job, taking care not to pull off any foam with the paper. Finally, bend the foam core and glue precut braces to fit the inside of the bend. If this bend (or sharp fold) is bumpy or creased instead of smooth, it may be that you are bending the paper along its fibers. Cut a new piece and try the bend acoss the fiber direction. If this doesn't work, the foam core may just have a poor quality paper. The finished curve, ready for glued braces, is seen in Figure 5-11.

The above descriptions should allow for many model possibilities. Foam core, of course, can be used together with graphics and in combination with other materials. One of its main disadvantages is the fragility of the material. Foam core models are difficult to handle and ship, and are usually used for temporary presentations. There is also the question of its flammability. Big foam core displays may be a fire hazard in a crowded hall, and in some places such use has been banned.

Fig. 5-10. Preparing the soft bend. Photo: Corinne Neuffer.

Fig. 5-11. A finished soft bend. Photo: Corinne Neuffer.

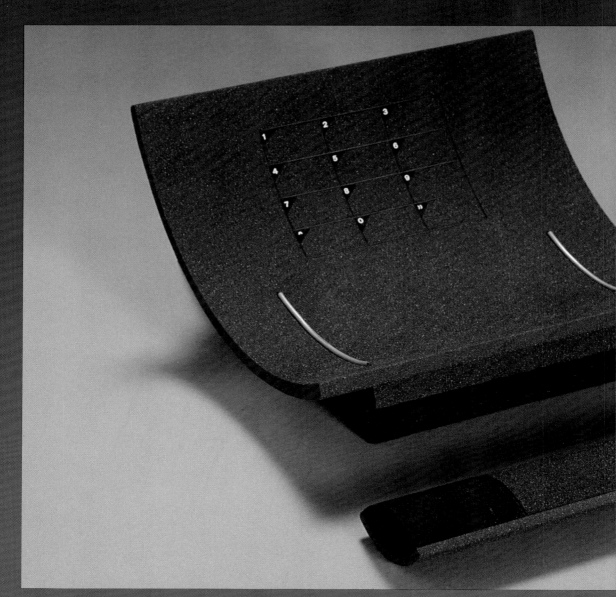

Photo, design and model: Caroline Sechehaye.

CHAPTER 6
FOAMS

OPEN AND CLOSED CELL FOAMS

There are many kinds of foam, but only a few are commonly used in modelmaking. These include only closed cell foam, as opposed to its opposite—open cell foam.

Foams such as bread, plaster, natural sponge and foam rubber have open cells which are connected to one another. Such foams absorb liquids into their entire volumes. Closed cell foams have cells that are distinct bubbles which are separate from one another. Liquids will only fill the cells that have been broken open at the surface.

Closed cell foams offer both an advantage and a disadvantage when used to make models. The advantage is that the model will accept coats of filler, primer and paint on its surface only, whereas an open cell foam would continue to absorb liquids until the entire model is saturated. This would cause real problems in drying and the cost of the liquids used would skyrocket. The drawback is that once the liquid dries on the surface of a closed cell foam model, it forms a hard shell only 1 or 2 mm. deep—as deep as the average bubbles broken open at the surface by shaping and sanding. Shipping, or even handling, while finishing the surface will easily cause this coating to break or chip off, hence an annoying repair job. Surfboard makers and their foam suppliers face this problem, and a possible solution is to punch the model's surface full of tiny holes, a procedure described later in this chapter.

Fig. 6.1. A finished foam model. Photo, design and model: Marcel Soiron.

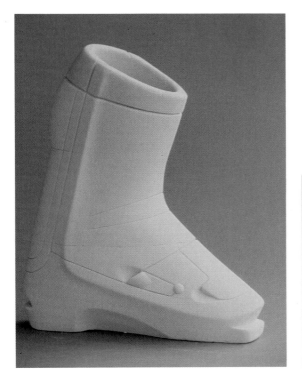

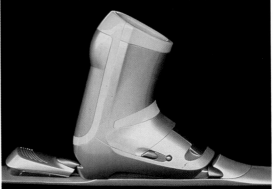

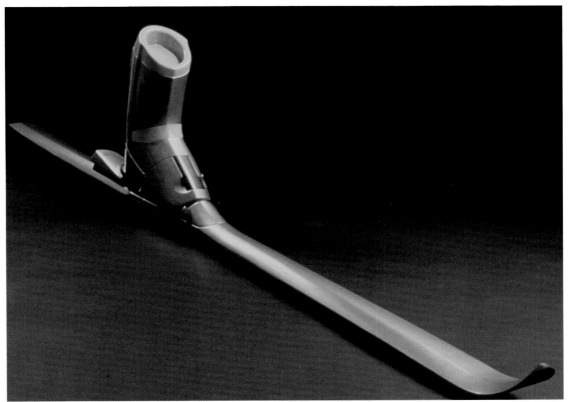

Figs. 6-2, 6-3 and 6-4. Unfinished, then finished urethane foam model. Photo, design and model: Ulrich Schraudolf.

MODELMAKING FOAMS

The two rigid closed cell foams most often used in modelmaking are polystyrene (also called styrene) and polyurethane (urethane) or isocyanate. PVC foam, available in sheets, is used for thin work, for example, where styrene and urethane would not be strong enough. All of these are available as buiding materials, since they are used as insulation, sandwich core constructions, and so forth.

It is often thought that styrene foam is used only for quick mock-ups and urethane foam for finished models, but there are tricks that make their uses interchangeable. Since urethane costs much more than styrene, this factor is to be considered when choosing materials for a project. The really important difference between the two foams is that styrene is a thermoplastic and urethane a thermoset (chapter 7 describes these differences, and how they can make for success, or failure, in even simple modelmaking operations). One former student forgot this difference. He took his styrene foam model to the spray booth, sprayed it with acrylic lacquer and watched as the whole thing melted down into a sticky mass.

Fig. 6-5. Tea set. Photo, design and model: Ulrich Schraudolf.

POLYSTYRENE FOAM

Usually white, sold mostly as insulation, fairly cheap and nontoxic (except for its fumes, when melted by a hot-wire cutter, for instance), this foam is often used for temporary displays and quick mock-ups of designs in three dimensions. It can be bought in various densities, ranging from about 15 to 40 kilograms per cubic meter. The lowest density (the "lightest") is the cheapest and most commonly available. It has big air (or CFC gas) bubbles. Because of this it is easy to shape, but the large, grainy structure leaves a very rough surface. This is what has given styrene the reputation of a "mock-up only" material. Higher densities have smaller bubbles and can be finished to finer surfaces. The following techniques are used to work styrene foam. Some of them are the same as for urethane foam, which will be discussed later in this chapter.

SHAPING

It is easy to shape styrene foam by using simple hand tools. Though it is not unpleasant to use, as is urethane foam, its lightweight grains cling to clothes and fly everywhere, making a real mess in the model shop. Saws, rasps, wire brushes and abrasive paper all do a quick job in shaping it. A hot-wire cutter may also be used, but its advantage is limited. The hot wire, acting as an electrical resistance, melts its way through styrene foam, thereby requiring good ventilation to disperse the fumes produced. The hot wire leaves a glazed surface along the cut line, which often has to be sanded off to provide anchorage for the finishing coats to be applied.

GLUING

Although styrene foam exists in big blocks and slabs, it is sometimes necessary to glue it. The project may be a very big model or a hollow construction of glued foam parts, or a repair or design change may be needed. Several points must be remembered when gluing.

Since styrene foam is a thermoplastic, it will not only melt with the heat of a hot-glue gun, but will dissolve in certain glues, such as ordinary neoprene contact cement or "dope" type glues in tubes. Special contact cement, containing a solvent that will not dissolve the styrene, may be used. Other glues, such as vinyl dispersion wood glue and epoxies, will also work without destroying the foam.

When gluing foam parts for later surface shaping and finishing, keep the glue inside the joint, away from the planned final surface. This is not always easy to plan, but it will prevent problems later—for example, when gluing pieces of foam (this also holds true for urethane) to construct a hollow model, for economy or lighter weight. A planned shape is shown in Figure 6-6. If the glue is everywhere, right up to the edges, the result of sanding the surface will be as shown in Figure 6-7. This is hard to correct.

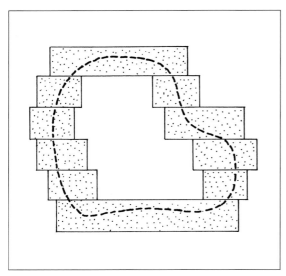

Fig. 6-6. Glue at edges.

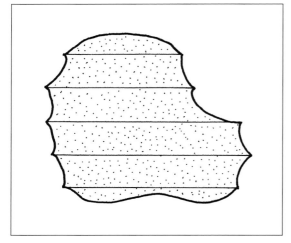

Fig. 6-7. The problem.

All dried glue sands differently from foam. Abrasive paper will wear away the foam more than the harder glue and will leave protruding ridges of glue along the joints. These must be cut away to give a continuous surface.

Gluing may also be needed for a repair or an "add-on" design change on the model. This can be done more easily with foams, both styrene and urethane, than styling clay enthusiasts would like to admit. It helps a lot if this is done before any hard crust surface filler is on the model, but it is still possible on surface-filled foam. Simply cut out the foam from the area where a repair or change is needed. A wedge-shaped cut is best, since it allows a good, tight fit of the wedge-shaped piece glued in. Again, keep the glue inside, away from the surface, to avoid sanding problems.

SURFACE FILLING STYRENE FOAM

With styrene foam, surface filling must take into account that styrene is a thermoplastic and is therefore soluble in many solvents. So many, in fact, that there are few fillers, primers and paints that can be applied directly to its surface. The most common products that will not harm styrene are:

- Vinyl dispersions in water (glues and paints).
- Shellac (but only with pure alcohol—with no acetone or aldehydes—as its solvent).
- Epoxies (paints and glues).
- Plaster and plaster-based fillers.

Once hard, some of these (epoxy, for example) will form insoluble barriers on the foam's surface which allow other, normally dissolving liquids to be applied without harming the surface. These "risky" liquids include normal neoprene cement, acrylic and cellulose lacquer solvents such as toluene, xylene, acetone and methyl chloride and polyester resin solvents.

A former student played with solubilities in model work in a mountain backpack model project. It was to be a rigid construction (it looked like a big, redesigned egg), but it had to be hollow, with a door to open it. There wasn't enough time for a casting and molding operation, so another approach had to be found. The solution was to first shape the model in solid styrene foam. Then it was coated with epoxy paint. Once the epoxy was dry, several coats of polyester resin were applied to its surface. This resin would have dissolved the foam model away if it were not for the epoxy coat. When that was dry, the door opening was cut out of this hard shell, and the piece was saved for the door, to be hinged on later. A can of acetone was then poured into the opening, which dissolved away the original styrene foam model inside, but not the hard shell around it. Reaching inside, more polyester resin was applied to the interior surface, reinforced with glass fibers for strength. This was much easier than fiberglassing the outside, which would have caused surfacing and sanding problems, since glass fiber is just too hard to sand away. Finally, the outside surface was rigid enough to be sanded and painted. The best use was made of not only the advantages, but of the "inconveniences" of thermoplastics and thermosets.

Foams are coated with fillers that not only dry hard for rigidity, but can be sanded much finer than the foam itself, as preparation for gloss painting.

Styrene foam may also be used for "bucks" (armatures) when building big models such as fifth-scale cars. This avoids slower, heavier work with wood or particle board. The most loved material for such finished car models is styling clay, which is heavy, fragile, expensive and almost impossible to paint well. Styrene bucks can help eliminate some of these clay problems, but there are precautions that need to be taken. First, the final surface of the model must be determined as closely as possible, as with other types of bucks. The buck should be built to allow $3/4$–$1^1/4$" (2 or 3 cm.) of clay to be laid on to create the final surface. Too thin a layer of clay might allow no room for safe surface work and design changes that require clay removal. Once the rough buck is built, its surface must be prepared for the application of the clay. Holes about $1/3$" (one centimeter) deep and in diameter are made all over the foam's surface to provide good anchorage for the clay layer. Then, since this clay is expensive and often recycled after the model's presentation (if it is clean and free of grit that makes smoothing a nightmare), the surface of the buck must be covered with a hard shell, like shellac, to protect the clay which must later be peeled off. Apply two or three coats to the foam, letting each coat dry completely. The result will be a fast, cheap, lightweight buck that will not contaminate the styling clay.

Fig. 6-8. Sanding filler on urethane foam. Photo: Yves Renaud. Design and model: Yves Renaud, Michael Pickstone and Isao Sasaki.

POLYURETHANE FOAM

This dust-producing thermoset foam is available in densities ranging from about 30 to 100 kilograms per cubic meter. Not many manufacturers sell the higher densities, since it is difficult to produce a dense, homogeneous product. The higher densities cost more and are harder to shape into models, but finer detail can be worked into the close-grained surface, and a dense foam model has the added benefit of being less fragile. Even coated with surface fillers, a low density foam may easily break. Polyurethane is so abrasive that it can sand itself: a small scrap of the foam (preferably of a lower density) can be used for final sanding, when the edges of abrasive paper might mark the surface. Urethane foam, even the higher densities, has little tensile strength, so it can easily break during model work. For this reason, the foam should be used for bulky models, not, for example, for airplane wings or other fine projections. Because it is a thermoset, urethane foam will not melt with heat and will not dissolve in most liquids. Therefore, it cannot be hot-wire cut, but can be directly coated with the common liquid or paste fillers, primers, glues and paints that would ruin styrene foam surfaces. This, plus the fact that it can be sanded finer than styrene foam, makes it a very popular material for design modelmaking, architectural maquettes, and television, motion picture and theater settings.

In some ways, that's unfortunate. Polyurethane is a crystal-hard substance, even as a foam, and with very little sanding and rasping, the dust finds its way everywhere and makes itself felt. This dust is rough and abrasive, and at the end of the day it is in the hair, ears, nose—everywhere. According to manufacturers, it is an "inert" substance (just like flour, coal, wood and stone dust which do cause recognized industrial illnesses, including silicosis). Polyurethane dust often causes skin irritation, especially under snug-fitting cuffs, collars, bras, trouser belts and socks. A good shower at the end of the day helps a lot, at least until going back

Fig. 6-9. Perfume bottles in hard urethane foam. Photo, design and models: Marietta Manos.

to the model shop the next day. Polyurethane dust loves fresh-painted surfaces, sticky styling clay models and delicate eyes. Should your eyes catch some dust, don't rub them. Just lean your head down for a few minutes and let the dust "cry" out. Remember to wear a good dust mask when working with this foam. As inert as it is claimed to be, it is already suspected of causing lung cancer. Many design studios, from Europe to Japan, forbid its usage.

Fig. 6-10. Urethane foam and the tools used to work it.

Fig. 6-11. Applying filler. Photos: Corinne Neufer.

SHAPING AND SANDING

To shape and sand urethane foam, the same hand tools as those used with styrene work well, with the exception of the wire brush. Power tools, and even heavy machines such as the bandsaw and planer are also used more on this foam than on styrene, mainly because it is more rigid and because it will not melt with the heat created by some high-speed machine work. Sanding should be done dry, mainly because final sanding will often be followed by surface filling, priming and painting. Figure 6-10 shows a block of urethane foam being prepared for model work, along with some of the typical tools used to shape it.

SURFACE FILLING AND REPAIRS

Before applying filler to the foam, the surface must be prepared. This means going over the entire surface with a fine nail or spiked roller to punch tiny holes throughout. This will provide anchorage for the filler and help prevent its being chipped off during later handling. If this surface crust breaks through to the foam once, there is a good chance that it will happen again before the model is finished. It is best to avoid all this by applying a thick coat of filler on a well-prepared surface, and by being careful while handling, sanding and storing the model. Figure 6-11 shows a small piece of urethane foam with its surface holes being coated with a paste filler. The filler should be applied by spreading it on the surface in several directions. The trick is to try not to trap air bubbles in the filler. After sanding the dried filler, bubbles would have to be filled and sanded again.

Practically any type of filler may be used on urethane foam without harming its surface. Avoid the one-component fillers which harden as their solvents evaporate, however. These are thermoplastics, and may dissolve when a coat of primer or paint is applied later. After many weeks of work and a lot of sanding of the surface filler, one student was nearly finished with his fifth-scale car model. In a hurry, he bought a spray can of cellulose-based filler. This is a thermoplastic, even once dry, and can dissolve in most primers and paints. It did just that, and the student had to wash the whole mess off with a solvent before painting it at the last minute for his presentation. He was at least lucky to have used urethane foam for the model, which survived the bath.

Polyester filler, made from the resin with a powder such as talc mixed in, is often used because it is tough, yet easy to sand because of the added powder. Pure polyester resin sets too hard for sanding. It may be used pure to provide more strength before applying a sandable filler, but the resin must be applied with care. If it drips down a surface and dries that way, the bumps will be hard to sand down. Apply thin coats, avoiding drips and resisting the temptation to finish the hardening mix each time if you mixed too much. Thixotrope may be added to the resin to prevent dripping, but it takes some practice to correctly mix and use this powder. Too much of it in the resin will be like trying to apply partly gelled resin, and the result will be the same bumpy surface you were trying to avoid by using it. In correct proportions, thixotrope gives a liquid the ability to be stirred and brushed onto a surface, then set again when left alone to a non-dripping gel. When mixing it with a resin, rather than try to predict the correct proportions, just keep adding the powder and mixing, stopping once in a while to let the mixture set for a few seconds. Then try pouring it. If it does pour, more thixotrope is needed. If it gels even while trying to mix it, there is already too much powder in the mix and you'll have to start again. Of course, such mixing takes a little time, so be sure to add catalyst to the resin only after the correct mix is ready. A catalyzing resin could be the cause of your gel.

If you do apply pure resin before a sandable filler, be sure to rough sand it and blow off the dust before applying the filler. Every care should be taken to ensure adhesion of applied coats at every stage.

MATT FINISH FOAM MODELS

With urethane foam (and, to a lesser extent, with dense styrene foam), it is possible to make use of the foam's naturally matt, velvety surface. The following points are worth considering when planning a model, or even a design to be presented as a foam model:

- The design itself may call for a matt surface, in which case it would be a waste of time to prepare for a gloss surface by filling and fine sanding, only to make the final surface matt again.
- A matt-painted foam model is not necessarily a quick mock-up.
- The time or budget may not allow for filling and sanding for a gloss finish, and this (often for light reflection studies) may not even be necessary for the model's presentation.
- Not only matt surfaces, but even fine details may be easier to produce without all the filler, which tends to bridge over and hide details when applied to foam, requiring them to be produced again in the dry filler coat.

The techniques are simple and include a few tricks to reduce work and ensure a continuous matt surface. Shape and finish the model down to its final dimensions (as opposed to leaving room for a few millimeters of filler to reach the final size). Give the surface finer details and a smoother final sanding than you would with a model to be coated. Do not apply any kind of surface filler or primer, not even for repairs to the surface. No patches of filler or primer will ever sand to the same texture as the foam, and the final thin coat of matt paint will not hide this difference. To repair pinholes, cracks or small defects in the foam's surface, use a variation of one of the methods shown in Figures 6-12 to 6-14. To repair a small hole in the surface, for example, use a sharpened pencil to make the hole bigger. Then use the pencil sharpener to make a matching point with a piece of the same foam as

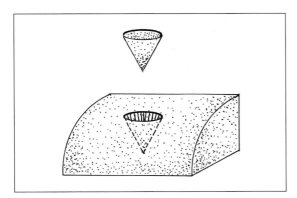

Fig. 6-12. Plugging small holes with foam.

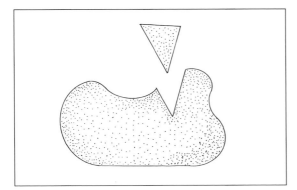

Fig. 6-13. Repairing damage.

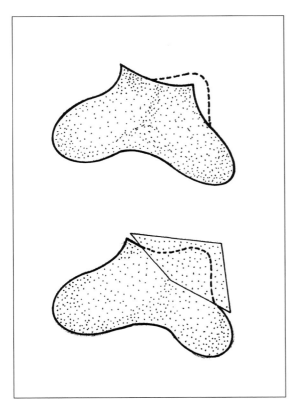

Fig. 6-14. Adding on for a design change in foam.

the surface. Drive this point into the hole, with only a drop of glue, applied only into the hole. When the protruding part of the foam point is sanded down to the surface, the repair will be invisible. Proceed in the same way for other types of repairs, cutting wedges of foam to fit wedge-shaped holes cut into the surface to be repaired. This same method, on a larger scale, can be used for add-on design changes in foam.

To finish a matt foam model, first blow the surface clean with compressed air. Then spray on one or two very thin coats of matt paint. Don't apply a primer first, and don't try for a gloss or eggshell finish. Masking is no problem for several colors, but follow the rules outlined in the chapter in this book on surface treatment and painting.

STORAGE AND SHIPPING

Foam models are easier to handle and transport than heavier models, like those made with styling clay, but foam is fragile enough to be easily damaged on impact. Foam models should be stored or shipped in sturdy crates, and protected inside with a thick layer of softer foam.

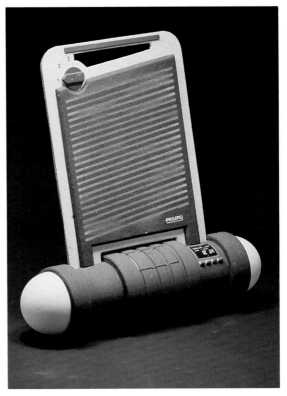

Fig. 6-15. A matt urethane foam model. Photo, design and model: James Hamilton.

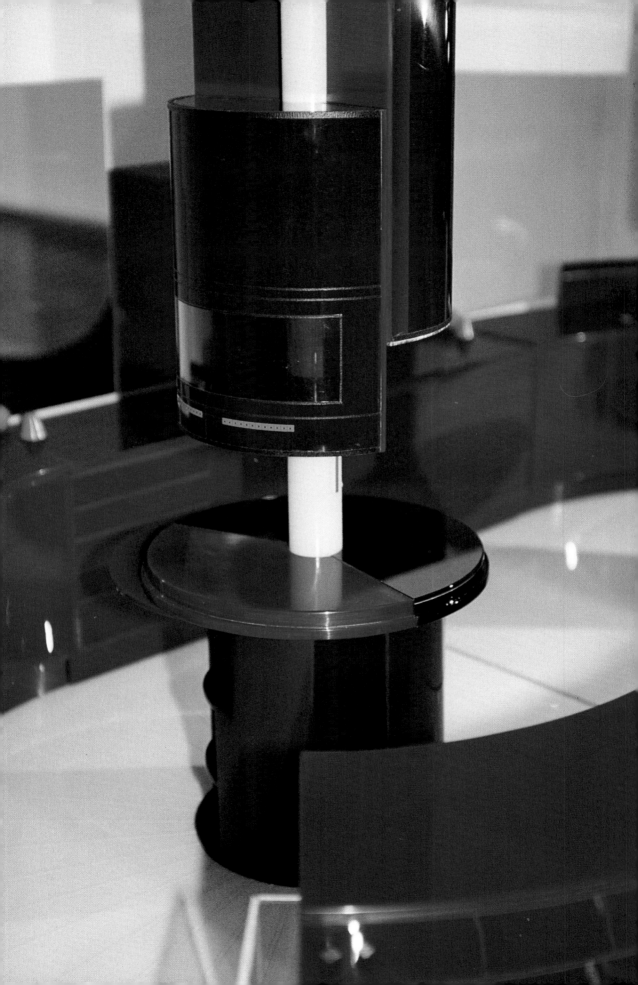

THERMOPLASTICS
AND
THERMOSETS

BASIC DIFFERENCES AND EXAMPLES

Don't be intimidated by the "technical" sound of this chapter's title. The fact is that the information presented here, along with some of the material from other chapters, actually shows some shortcuts to help bypass the confusion that often arises with modelmaking materials. Rather than spending a long time making and repeating mistakes, this short chapter can show how to avoid many problems and ruined projects simply by understanding what can and cannot be done with plastics.

Thousands of resins and solid plastics are produced today. A technical service engineer from a big American company recently admitted in a speech that he knew less about plastics today than he did thirty years ago when he started his job. Fortunately for the design modelmaker, the choice is reduced to just a few, either in the form of liquid resins or solid plastics. But there is a small amount of chemistry and physics involved in the work which may be useful to learn.

The two basic categories of natural or synthetic plastics are:
- Thermoplastics, formed as liquids at relatively high temperatures. The liquids cool to solid plastics (in the form of a sheet, tube block, etc.) at room temperature.
- Thermosets, or thermosetting resins, set to a solid state by a chemical reaction.

Both of these types may exist in either solid or liquid form.

The few, but very important differences between the two are:
- Thermoplastics can be melted back to a liquid when heated. They are also soluble in specific solvents.
- Thermosets, once hardened, will not soften and melt with heat. They are insoluble in most solvents.

Fig. 7-1. Polystyrene foam dissolved by a glue with the wrong solvent. Photo: Monique Carroll.

In design modelmaking, as well as in many industrial applications, there are important differences in the way these plastics may or may not be used. The following table shows examples of the most common plastics, as well as examples of the differences between them which often affect modelmaking.

THERMOPLASTICS	THERMOSETS
Styrene, acrylic, acetates, PVC, polyethylene, vinyl, polycarbonate: as solids, paints, glues, etc.	Epoxy, polyester, polyurethane: as 2-component liquid systems, solids, glues, paints, etc.
Soften and melt with heat	Will not melt
Dissolve in solvents	Insoluble
Can be thermoformed	Cannot be thermoformed
Can't be used as molds in thermoforming	Can be used as molds in thermoforming
Can be solvent bonded	Can't be solvent bonded
Do not resist cleaning, painting, etc., with all solvents	Safe to clean and coat
Can be hot-wire cut (foams)	Can't be hot-wire cut
Can be solvent-cleaned off tools, spray guns, etc.	Almost impossible to clean once hard
Solid surface can be painted, etc., with liquids that dissolve (etch) the surface, but evaporate fast enough to do no harm; such paints adhere well	Can't be etched; must be rough sanded for good adhesion
Machine sanding, polishing, etc., can melt the plastic	Will not melt
May dissolve in certain glues, fillers, primers and paints	Safe to coat with any liquid
Glues and fillers may smear with machine sanding friction heat	No problem sanding

The information presented in the chart relates to the strengths and weaknesses of both thermoplastics and thermosets. In the actual practice of modelmaking, these may be interchangeable: what is normally taken as a drawback may serve a useful advantage, or vice versa. A basic understanding of plastics and some imagination can lead to exciting and original projects, as well as an increasing control over past mistakes. The following stories, all true, may help to illustrate this point.

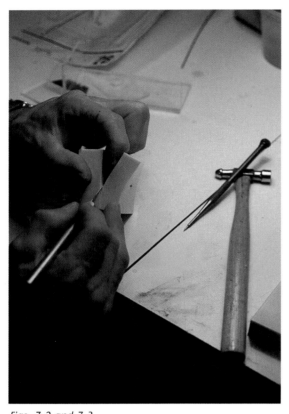

- Gwendoline spent weeks shaping and sanding a full-scale model of a ski design. She used styrene, solvent bonded with acetone (OK, so far) to laminate the thicker sections. During the final phase, she used polyester filler for some surface defects. The result: the styrene ski dissolved and softened in the filler's solvent, requiring a long wait and a big repair job.
- Paul finished his scale model of a boat design and primed the surface with a spray can of one-component primer (a thermoplastic which dries as its solvent evaporates). Then he sprayed it with a polyester topcoat whose solvent dissolved the primer into a sitcky mass of blisters. The whole model (in urethane foam, happily) had to be washed down in solvent and left to dry.
- John was running out of time and decided to use his original model (in styling clay, a waxy thermoplastic) as a mold to thermoform a transparent top for his car model. It would have been safer to use the original to make a plaster or fiberglassed resin mold, but it was a quarter-scale model, and that would have been a big job. Of course, the strong heat of the softened acrylic sheet thermoformed

Figs. 7-2 and 7-3. Slicing open an RTV rubber mold, into which urethane was poured to make a rubberized foot for an electronic device. Photos: Micaela Porta. Design and model: Alex di Nota. Reproduced by courtesy of Henry Dreyfuss Associates.

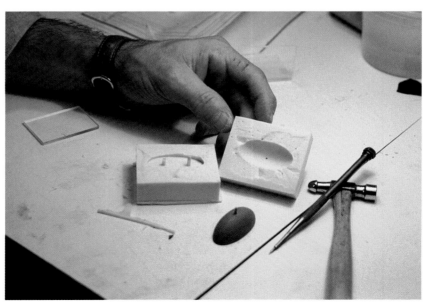

against the clay melted it down to something like chocolate pudding—but the acrylic cooled and set hard just in time, and somehow the very difficult vacuum pull worked at the first try. He was lucky, and really happy.

- Anna tried for almost an hour to hot-wire cut a piece of urethane foam. She did make a dent in it, but also bent the wire out of shape trying.
- Mitchel used a thermoplastic primer on his vacuum forming mold. The mold bubbled up with the heat, creating a design he hadn't even dreamed of.
- Heike's maquette of a beauty salon interior was made of PVC foam sheet and was extremely dirty with pencil and marker smudges. This foam is hard to sand clean, especially near tight corners. So she tried solvents, starting with the safest and venturing as far as methyl chloride for some of the tougher dirt. She tried everything, except cyclohexane, which would have dissolved all her work.
- Christine's model of a perfume bottle, in clear-cast polyester resin, was very good, but she wanted it to be perfect. She had drilled a hole into it to insert the polished aluminum perfume dipper, and the scraping action of the drill bit left the inside of the hole matt, thereby obscuring the dipper. With acrylic, she would have been able to quickly shine the inside of the hole with some methyl chloride. But the polyester would not dissolve. It was also too hard to try to polish inside the tiny hole, especially since polyester resists abrasion. She poured in a little fast-drying clear lacquer and then poured it out. Once dry, it left just enough of a shiny film inside to fill the matt scratches, and her dipper was visible.

These accidents—and occasional bouts of good fortune—are all too common: styrene cups filled with solvent which don't make it across the room; acetate sweeps being ruined in hot car trunks or next to radiators; the wrong contact cement used to glue styrene foam. It isn't necessary to remember all these stories; surely you can add to them with stories of your own, or of your friends. As usual, take the time to test new or unfamiliar materials and ideas on scraps, or at least on hidden parts of a model. Starting ruined projects over is time-consuming and frustrating. Read the labels on cans of paint, glue, fillers and other materials, and use them only as thermoplastics and thermosets may be used. This sometimes requires planning and often forces a choice to be made concerning materials at one or more stages of the model work.

Photo: Micaela Porta.

CHAPTER 8

STYLING CLAY

CHOOSING CLAY

Styling clay, well-loved by car designers, is probably the strangest material used in modelmaking today. It smells awful because most types contain sulfur, is heavy and slow to work, makes fragile models and is almost impossible to paint well because of its greasy, waxy surface. It is also expensive, and time-consuming recycling of the material from models is often necessary. In hot weather, styling clay may crack on a model surface, and yet it requires controlled heating to make it workable, in clean rooms where no dust is produced from other work. Its sulfur content also inhibits some necessary chemical reactions, such as silicone rubber hardening. And it is a thermoplastic material that not only melts with heat but dissolves in many solvents. However, it is not toxic and can be worked to a slick finish, even without painting.

For anyone who still wants to use clay, the following information deals with the basic approaches to the work (some of the techniques are also useful for shaping other materials). For advanced clay styling, it is almost always necessary to work side by side with an experienced instructor.

THE MATERIAL

Most brands of styling clay are only firm at room temperature. The clay is tooled at room temperature, but must be heated to a softer consistency for the initial work of laying it on. Heating is done in a temperature-controlled oven (not one used for food preparation), usually at about 122–140°F (50–60°C). Too cold and the clay will not be soft and malleable enough for the initial laying on. Too hot, it will decompose into a gritty mixture that is impossible to work, can contaminate good clay and must be thrown away. Special ovens built for use with styling clays are not easy to find. One could substitute with a discarded refrigerator or deep freezer, both of which provide good insulation. Use a strong light bulb inside as a heat source, along with a thermometer to check the temperature.

CLEANLINESS

Whether warm or cold, styling clay is sticky and can quickly become contaminated with grit that makes it impossible to smooth final surfaces. It is important, then, that the work area, table surfaces, tools and the buck stay clean. No dust from other model work should be produced in the clay workroom and the process of making the buck and final painting (if it is attempted) should include the possibility of recycling clean clay. Once the clay model is started—and it may take weeks to finish it—it makes sense to cover it with light polyethylene sheet at the end of each work period.

BUILDING BUCKS

The buck is usually built to provide a sturdy support for big clay models (such as fifth-scale car designs) and to economize on the amount of

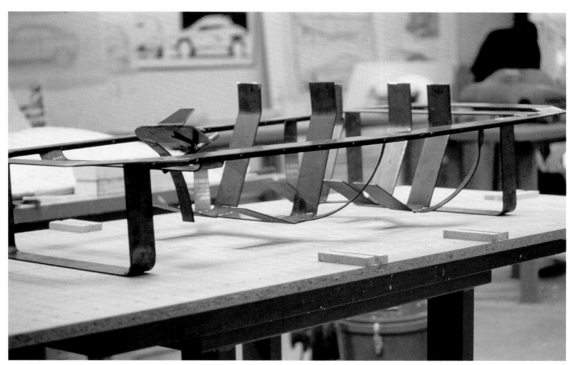

Fig. 8-1. A clay buck and armature. Photo, design and model: James Dean.

clay needed for the work. The buck should be built with the possibility of design changes in mind. Following scale drawings, the rough, step-shaped buck should allow for about 1¹/₄" (3 cm.) of clay to be laid on for initial work, or more on parts where there is already a question of design changes. This will permit the removal or addition of clay without contamination by digging away at the buck material inside. An armature for such a buck is shown in Figure 8-1. In the rear, a styrene foam buck is being prepared for clay.

Once the buck is built, it should be drilled full of holes—about one centimeter in depth and diameter—to provide anchorage for the clay. Whether the buck is made of styrene foam, particle board or plywood, it should be coated with several coats of shellac and allowed to dry completely. This sealing will help keep crumbs of the buck material from being rubbed into the clay while laying it on or removing it.

LAYING ON CLAY

The clay should be soft and malleable when removed from the oven for easier work and better anchorage to the buck. As a handful of clay cools and hardens during the work, it should be returned to the oven and exchanged for warm clay. If a new day's work requires adding more clay to a cold clay surface, this surface should also be gouged with holes to allow the new clay to hold better.

Press the clay on the buck (or on clay from the previous day) by rubbing it along the surface as it is added. If it is only lightly pressed to the surface it will not stick well. Cover the entire surface of the buck, without trying for a smooth surface or any detail. Try to rub out trapped air bubbles and to build up a layer of clay that will be ³/₄–1¹/₄" (2 or 3 cm.) thicker than the final surface needed. When enough clay is on, let it cool to room temperature and protect it from dust and grit by covering it with a clean, plastic sheet.

Fig. 8-2. Pressing warm, soft clay onto a buck. Photo: Monique Carroll.

Fig. 8-3. Scraping cool clay with a special clay tool. Photo: Monique Carroll.

CLAY TOOLS

Styling clay tools are quite diverse, developed in all shapes and sizes by good clay modelers. Some of the common tools available for use with other types of modeling clay may be used, but the waxy styling clay has its own needs and better work can be done with tools tailor-made for it. Hard to find in many countries, these tools can be very expensive. Chapter 14 shows how to get around these problems by making some of your own, which can be as good as anything available in stores.

In addition to the shaping tools, some special measuring and referencing equipment is also needed:

- A squared grid board, with the grid marked to match a gridded top view drawing of the model. The clay model will be built directly on this board. Some more sophisticated boards are made with grooves or sliders along each edge for sliding reference pointers.
- Male and female templates, made of styrene or acrylic plastic sheet, traced from section views of the drawing. Templates may be used for scraping as well as pressing guide marks into the clay.
- A heavy square (hardwood or metal), with edges wide enough ($2^1/_2$–$3^1/_4$", or 6–8 cm.) to allow a flat pointer to slide flatly along its surfaces without rocking. This square is positioned along the grid board for marking surfaces with the pointer.
- A flat pointer, in wood or metal, used against the square at various intervals along the grid board. The pointer scrapes vertical reference lines into the clay surface.
- A trusquin, or adjustable pointer, mounted on a heavy, flat base. Limited by a straight or contoured rail along the model, it is used for marking heights and depths on the clay surface.
- Steel rulers, for measuring within grids and other checking.

SHAPING CLAY MODELS

Shaping styling clay is the point at which experience comes in handy. There are lots of tricks of the trade to be learned and discovered, developed over years of practice by individual modelers.

Once enough clay is on the buck, it must be marked with reference guides for the final intended surface using the tools described above. The spacing of the section marks depends on your eye and experience, as well as on the surface curvature. Tighter and more varying curves need closer spacing for control of the continuity of the curves. This is true for shaping all materials, from wood turning to foam or clay models.

The positions of the imprints were previously marked on the rough model by a square and flat pointer used along the grid board. A trusquin could also be used to score horizontal marks along the model, or to mark identical areas on each side of it for symmetry. In addition, the trusquin could be used to score or imprint the clay for referencing instead of using a template. Here too, the decision depends on the precision desired, experience and the relative ease of one of the methods for the particular job.

With the reference marks in the clay (they may have to be made anew during shaping work) the model is ready to have some clay removed, a crucial step closer to the final surface design. The tools used are mostly forms of scrapers and cutters, a little bit like miniature garden tools. Knives, or scrapers with teeth, are used for initial work, removing a lot of clay at a time. The smooth-edged tools are for more precise work and some are shaped to produce concaves in the surface. Spring steel bands, in varying flexibilities and sanded fine along the narrow edges, are used for smoothing final flats, convex surfaces and slight concaves. If the clay is not going to be painted but must shine for highlight studies, thin acetate sheet can be used to "slick" or burnish the surface once the final form is obtained.

Using the tools, the clay is gradually removed between the imprint marks. It is slow work, requiring a "feel" for the way the clay behaves. Once in a while, clay may have to be added to build up a surface scraped too deep. Probably the most important technique in surfacing clay is to avoid making and riding on surface waves. Always pulling the clay tool in the same direction tends to ride along the wavy surface of the rough, laid-on clay. The clay will be scraped away, but the waves will remain as the work goes deeper. Figure 8-3 shows the tool being pulled in a direction perpendicular to the waves. It bridges over the high points of the waves, paring them down to a more continuous flat or curved surface. Both directions, or rather, all directions, are needed for this scraping action, since the bumps or waves can go in any direction.

Continue scraping, checking and adding clay if necessary as the whole surface approaches the drawing's design objective. Don't attempt any smoothing until the entire surface is evenly shaped with the roughing tools. Then, with the finish tools and flexible steels, continue crossing the diminished waves—using good, reflecting light—allowing the steels and scrapers to follow the contours which were previously roughed out.

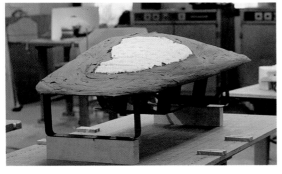

Fig. 8-4. Rough clay on a styrene foam buck. Photo, design and model: James Dean.

Fig. 8-5. Clay models in various stages of completion. Photo: James Dean.

Again, it is long, slow work, requiring the occasional addition of clay even for finish work. Should you wish to paint the finished surface, don't try to achieve fine detail or cut, engraved lines; they will only be hidden later by the necessary coats of sealer, filler, primer and paint to be applied. Slicking the model surface with acetate sheet instead would yield a shiny, brown surface.

SURFACE PREPARATION AND PAINTING

Clay is not easy to paint. It is full of waxes and oils, substances that any good painter will tell you must be removed from a surface before painting it. This is, of course, impossible with styling clay.

There are, however, several methods for getting the clay model past the presentation stage without peeling or blistering. Some of them work fairly well, and others are so difficult, even in terms of finding the necessary products, that they are usually not worth the trouble. The following description of a process that works must be followed closely. Other information, such as that contained in the chapter of this book on painting, should be used to increase your probability of success.

The finished clay surface is first wiped very lightly with a soft cloth dampened with alcohol. Then let the surface dry completely. It may take several hours because the alcohol may dissolve into the clay surface, which slows its evaporation. Then spray on two or three coats of good, alcohol-based shellac, allowing each coat to dry completely. Low-quality shellacs contain acetones and aldehydes in their alcohol, which will too easily dissolve materials in the clay (as they do with styrene foam). Make sure to run tests on scrap clay. Shellac sealer is used to prevent the clay from dissolving in the fillers, primers and paints to be applied later, all of which may contain solvents that can attack and soften the clay.

Once the shellac is dry, the clay is protected. It can then be sprayed with any of the favorite choices of paint systems: cellulose or acrylic lacquers, polyesters, etc.. Masking, especially tape pulling, requires special care on clay. Nothing sticks well to clay, and the dry paint may come off on the sticky tape. Polishing out paint defects, such as orange peel, is a delicate procedure. Clay is not a hard material, and too much pressure will squeeze it out of shape, cracking or breaking the harder paint coat.

HANDLING AND SHIPPING

The rule for clay model presentations should be "do not touch." If shipping and handling are necessary, extra care must go into crating the clay model, preferably in a sturdy box with the word "fragile" written all over it. The interior padding should be thick and very soft to prevent shocks. Repairing shipping damage on clay models just before presentation is no easy task.

FURTHER REFERENCES

Once again, there is much to learn and practice in order to become proficient in clay modeling. While literature is available on advanced work, it is best to have the privilege of working with one of the few clay experts in the design industry. Some of them have written about their work and others teach clay modeling in design schools around the world. Most of them work on a freelance basis or are employed by automobile designers.

Fig. 8-7. A finished clay model. Photo, design and model: James Dean.

Photo, design and model: James Hamilton.

CHAPTER 9

SOLVENT BONDING

SOLUBILITY OF MATERIALS

Thermoplastic sheet, tube or rod can be glued to construct models, but such joints are often weak. Much stronger joints are made by solvent bonding the parts together. This takes advantage of the solubility of thermoplastics in specific solvents. The following table shows this with a few examples of solids and appropriate solvents.

SOLUBILITIES OF SOME MATERIALS

Solid	Solvent
salt, sugar	water
acetate plastics	acetone, methyl chloride
acrylic plastics	methyl chloride
polystyrene	acetone, methyl chloride
PVC	cyclohexane
polyester, epoxy, etc.	none

Solvent bonded joints are strong because the material dissolves in the solvent. When the solvent evaporates from the joint, the two joined parts have become one continuous piece of the plastic used.

Fig. 9-1 and 9-2. A model of a flashlight design in acrylic. Some of its parts were solvent bonded to a precise fit to allow other parts to turn and slide. Photo, design and model: Caroline Sechehaye.

TECHNIQUES

There are just a few simple techniques used in solvent bonding. The choice depends on the size of the work, the finish required, joint positioning limits, strength and the speed of work desired. In any case, care is needed to avoid getting solvent on clean surfaces other than the joint. This would result in dissolved or "etched" spots and can sometimes be very difficult to correct.

Fig. 9-3. Dipping in an aluminum foil container. Photo: Corinne Neufer.

DIPPING

Dipping parts in a solvent is often used for smaller-sized work and produces a clean, strong bond. The solvent is poured into a shallow container to a level of 1 or 2 mm.. If the liquid is too deep, etched marks away from the joint will be visible later. A container can be quickly made by folding aluminum foil around the edge to be dipped, and then by slightly enlarging the opening in the foil. This saves solvent by reducing the amount that evaporates into the air. Work fairly quickly, closing the solvent container when not in use and returning unused solvent to the container after dipping. Many solvents are toxic when breathed and some are explosive, so use good ventilation, common sense and follow all the safety rules you can find.

Dip one edge into the solvent and leave it for ten to thirty seconds. Some solvents act much faster than others, so try tests with scrap pieces. The dipped edge should begin to dissolve and soften.

Press the soft edge against its joining part, using a square or angle guide to position it, and hold the parts together for about ten seconds. The wet edge will transfer some solvent to the dry part, which then begins to dissolve. The solvent begins to evaporate and the joint will set slightly. It will take an hour or so to develop strength as the solvent continues to evaporate. Leave it even longer before any heavy handling, sanding or the like.

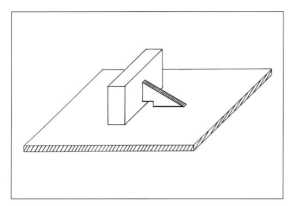

Fig. 9-4. Positioning parts with a notched guide.

Fig. 9-5. Bonding a very rough PVC edge with purchased PVC dope. Photo: Monique Carroll.

Since capillary action will draw the solvent into tiny spaces (for example, between a clean surface and a gluing guide), cut a notch out of the guide before using it. This will prevent solvent marks and will keep the guide from sticking to the joined parts.

CAPILLARY BONDING

In some cases, dipping is not convenient. With long parts, for example, or when the construction is already partly together, another technique—using capillary action of liquids—may be easier. The parts to be joined are positioned together dry. Careful, flat sanding of the butt edge is necessary before joining, since capillary action will draw the solvent only into the tiny spaces of tight fitting parts. Jagged, rounded or poor fitting edges will stay dry and a weak joint will result. Refer to the chapters on sheet materials, and on sanding and finishing in this book, for the sanding techniques used on sheet plastics.

A pipette, syringe or fine paint brush is used to apply the solvent to the positioned parts. Draw the solvent into the pipette, syringe or brush and carefully move the instrument along the joining edges. The solvent will be drawn into the joint and begin to dissolve both mating parts. Repeat this action several times, taking care not to drip solvent on clean surfaces. The result, once dry, will be a clean and reasonably strong joint.

DOPE BONDING

Despite the more recent definition, dope is still what it was in the good old days, when kids still used it on airplane models. A solution of plastic being bonded in an appropriate solvent, dope can be bought in squeeze tubes or it can be made very quickly. If you buy it, be sure to accept only a tube labelled for the specific plastic you want to bond. A tube of PVC glue will not work with acrylic plastic, for example, and acrylic glue will not bond PVC. The solvent in the tube must dissolve the plastic you are bonding. To make it yourself, just file or rasp scraps of the plastic into a little solvent and let it dissolve into a sticky, viscous solution. Use a metal or glass container rather than a plastic one, which just might dissolve while you are working. Keep this container tightly closed when not in use. The dope is applied along one joining edge with a brush or spatula and the two parts are immediately pressed together and held in place for several hours to dry.

Dope dries slower than the pure solvent used in other bonding techniques. The final joint is fairly strong, but not always clean. The big advantage of dope is that it is thick and will fill big spaces in joints, and can be used when edges are rough or rounded. Therefore, it is useful as a surface filler or for repairs to poor joints or damaged surfaces.

With any of the above techniques, if a poor fit is made, break apart the joint before maximum bonding sets in and start again. If there are solvent marks on a clean surface, fine sanding and even polishing may be necessary, which is not an easy job on transparent surfaces.

RELATED WORK

The information presented above can relate to other modelmaking work, as may be seen by reading the chapters in this book on painting, sanding, foams, and thermoplastics vs. thermosets. Keeping those subjects in mind—and lots of practice—will make you a better model maker. A few examples:

- The paint required for a certain plastic should contain a solvent that just begins to dissolve the plastic surface before evaporating quickly. The result will be a strongly adhering coat.
- Plastics should not be cleaned with the solvents that will dissolve them. A special exception is "dissolving" scratches or creating a matted surface.
- Surface filling, gluing or painting may ruin a nearly completed model by dissolving it.
- You may want to dissolve a plastic or foam part to achieve a special effect.
- Acetone, which was always sold as nail polish remover, seems to have been replaced by another solvent. If you buy this remover, you may dip away all day without bonding your styrene.

Increasing your familiarity with solubilities can help protect your equipment and yourself. It is not unusual to see a styrene cup, used to carry methyl chloride, not make it across the room. The same liquid can ruin plastic paint brushes and acrylic eyeglasses and watches. Most important of all is that many solvents (not yet water, everywhere, hopefully) are poisonous; they can "dissolve" vital compounds in the body if they are breathed or swallowed. Many are being banned for sale in various countries, which doesn't help modelmaking, but it is good news for our health.

Fig. 9-6. A styrene cup that lost its bottom and its contents, methyl chloride. Photo: Monique Carroll.

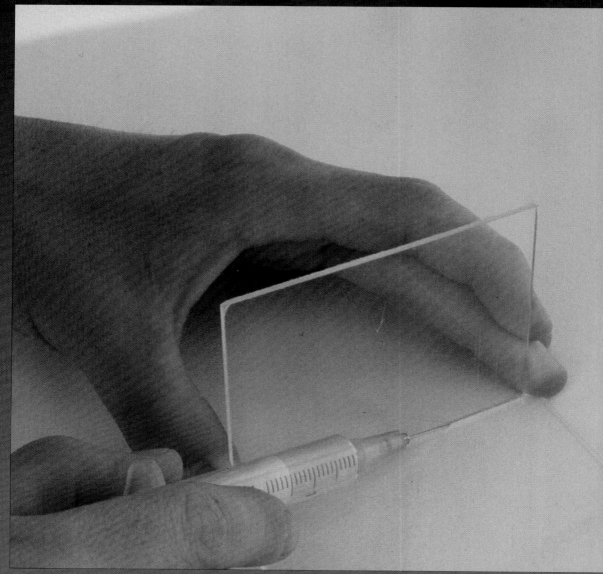

Photo: Corinne Neuffer.

CHAPTER 10

GLUES AND CEMENTS

SURFACE PREPARATION

In general, for any type of glue, both mating surfaces must be fairly tight-fitting and clean. Dust, grease or water must be removed, since these can fill the pores or cover the surface of the material and prevent the glue from wetting the surface for a good contact and a strong joint. Smooth, especially nonporous materials, should be sanded rough for a better grip (roughing increases the surface open to the glue). Poor-fitting joints should be made to fit tight, but only hand-tight, not requiring a hammer or mallet to force the parts together (this would force out too much glue from the joint). The glue should always be applied to both surfaces, since the spreading of the glue wets a surface better than just pressing glue from another surface.

Fig. 10-1. Roughing a smooth planed board edge for better glue strength. Photo: Monique Carroll.

CLAMPING

With many glues, especially on porous materials, it is necessary to press or clamp the surfaces together while the glue dries. The exceptions are with neoprene contact cement, which only requires a quick press or hammering together, and with epoxy glue on nonporous surfaces, which must only be held in place for drying. Clamping in the case of wood, with sliding joints such as the mortise and tenon, must be well planned and done quickly since many wood glues can seize parts together if they fit tight, even before they are in their correct position.

For most clamping, though, the work is easier. If not enough clamps are available, as often happens, or if the parts to be glued together have odd shapes that would cause the clamps to slip off, other means of pressing may be used. These can include string, tape, welding clips, clothespins or a heavy bag of sand that will fit around a contoured part. In most cases, a lot of pressure by clamping is not needed. In fact, excessive pressure may force so much glue out from between the parts that nothing is left for a strong joint.

Fig. 10-2. Clamping wood glue joints. Photo: Monique Carroll.

TYPES OF GLUES AND CEMENTS

WOOD GLUE

Usually water-based wood glues are inexpensive and strong on all porous materials, including paper and foams. Various types are available, for example the common vinyl glue. Some are waterproof (polyurethane) and some are powders that must be mixed with water. Glued joints must be pressed or clamped until dry, usually from three to eight hours, at a minimum shop temperature of 60.4°F (15°C). These glues will hold plastics and other nonporous materials slightly, but such surfaces should be roughed before gluing. One big inconvenience of white vinyl glue is that it is a thermoplastic and softens with even a little heat. When machine sanding such glued joints, the friction heat will smear the glue along the surface and quickly clog up the abrasive paper. One advantage is that many varieties dry transparently. To patch or fill a wood surface, a little dust from the same wood mixed with white glue will make a better filler (for the color and adhesion needed) than anything else available on the market. The smearing problem remains, however, so it is best to scrape off the excess before it dries for sanding.

HOT-MELT GLUE

Working with an electric hot-melt glue gun is fast and easy, but it does have its limitations. It works well with paper, some foam core, urethane foam and other porous materials, but does not always give a strong bond. Since most hot glue guns squeeze out too much glue at one time, it is also difficult to use it sparingly for small work. Remember, too, that its heat will melt such materials as styrene foam, which may be the foam inside some foam cores. Sticks or blocks of it may be used as a temporary quick fix, and hot glue can even be an inexpensive substitute for a flexible molding material instead of silicone rubber.

SPRAY CAN ADHESIVES

These provide a quick fix, mostly for paper work, but results may be mixed. When spray adhesives are supposed to stick, they don't always, and drawings may fall off the wall during presentations. Then the walls become impossible to clean—with any solvent—and must be painted. Spraying small work often takes place in a hurry, against clean tables, windows or carpets, which soon show the stubborn outline of the work. The toxicity of spray adhesives alone should probably make them banned from use everywhere.

THERMOSETTING RESINS

Epoxies, polyesters, polyurethane and cyanolite offer a special advantage: they are thermosets, and will not melt with heat or dissolve in most solvents. This makes them very useful for filling as well as gluing, when machine sanding heat or further chemical treatment (solvent wipes, priming, painting, etc.) would cause problems with thermoplastic glues.

They are usually very strong, but are less effective with some materials. Nonporous surfaces should always be sanded rough before gluing, and they should be tested on scraps of the material to see if they will dissolve it in case it is a thermoplastic. Each type also has specific advantages. Epoxies, for example, are the best for gluing solid (cast) urethane, and will not dissolve styrene. Transparent polyesters, mixed with opaque or transparent colors, are the best for color-matching fillers. Cyanolite, reacting with the humidity in the air, is extremely fast, but can be dangerous if used carelessly. Epoxies are poisonous, no matter what the manufacturers write on the labels.

CONTACT CEMENTS

Neoprene rubber cement is surprisingly strong and works quickly in many operations. It is thermoplastic, so the usual advantages and disadvantages apply. The gluing technique is simple, but critical. Cement is applied to both surfaces and allowed to dry completely. When the surfaces are not sticky if touched, they are ready for contact. Several coats will give greater strength, especially on porous surfaces. Once dry to the touch, the two surfaces are lined up and pressed or hammered together, and the job is finished—and so are the materials, if the surfaces were not lined up perfectly before contact. It is usually impossible to separate them without breaking should they need realignment. Concerning solubilities, ordinary neoprene cement does such a good job on polystyrene (solid or foam) that you might not even see the result after a few minutes. Styrene dissolves in the cement's solvent. A special contact cement, with a safe solvent, is available for gluing such thermoplastics.

SPECIAL APPLICATIONS

For gluing very big surfaces, when economy or health concerns are important, there are still the old-time glues, especially useful for porous materials. Cellulose wallpaper glue and simple flour and water are examples. Animal hide and bone glue are still available, but they don't smell as good and are more difficult to use. Such glues may be the best choice for big indoor projects such as papier-mâché work.

Specific solvents, for solvent bonding, are not really "glues." When used correctly, they are usually much stronger than any glue. Chapter 9 describes this special use of solvents for thermoplastics.

As mentioned above, most glues can be mixed with powders (as long as they are insoluble) to make fillers, match colors, or even as paint substitutes. As usual, test new ideas on scraps rather than risk damaging the model. Also, test new powders with polyester resin. Some powders have chemicals that inhibit the hardening of the resin, and this may cause a real mess.

TEMPORARY GLUING

For a temporary holding job using thermoplastic glues, heat is usually used to melt surfaces apart when necessary. Friction heat, hot air guns and hot water are most commonly used for this purpose. Heavy paper may also be glued between the surfaces, then split through its fibers to open the joint later. Such temporary joints are often used to hold abrasive paper on disks (better than any system available), for lathe face plate turning or to turn halves or quarters between centers and to hold foam parts for certain operations, such as a safe flat bottom for band sawing.

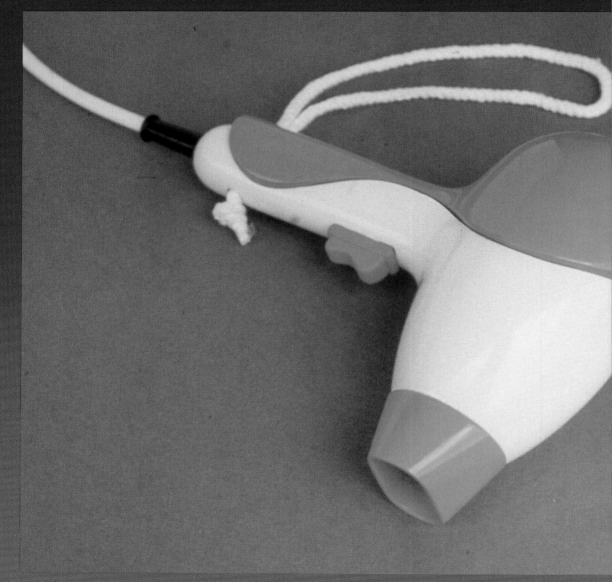

Photo, design and model: Ulrich Schraudolf.

CHAPTER 11

CASTING AND MOLDING

WHEN IT IS NEEDED

There are several reasons for making casts of models. First, multiple identical reproductions of models or parts may be needed, and time and precision can be gained by casting. Second, a hard and resistant model, in a material such as fiberglassed resin, may be needed, but it is easier and faster to make a master model out of a soft material like clay. Third, if a transparent model is desired, and the shaping of available transparent materials takes too long, or thermoforming cannot produce the designed shape.

Many materials may be used for master models. If a hard, nonporous material is chosen, the shaping work will be slower, but the surface sealing and mold releasing will be easier than with softer or more porous materials. There should be enough density and resistance in the material to accept details if they are a part of the design and to withstand the stresses of handling and releasing the mold, especially if this mold is to be made of a rigid material. The chapter in this book on materials and techniques may help you in making a choice.

Fig. 11-1. Men's toilet accessories made by casting. Photo, design and model: Marc-Antoine de Murault.

MOLDMAKING

Casting reproductions into molds in such fields as industry, fine arts and medecine covers a wide spectrum, making use of many materials and techniques developed over centuries. Some of the ancient techniques, such as metal casting into sand molds, are still used today. There are also fairly recent materials, such as silicone rubber, dental molding gels and resins, which have greatly added to the possibilities of the process. This chapter will present the materials and techniques most often used for casting and molding design models.

Molds may be made of many materials which set from liquids to solids around master models. Relatively few materials are used for design model molding, the most common being plaster, polyester resin and silicone (or RTV, for Room Temperature Vulcanized) rubber. Others, such as hot-melt glue, latex rubber and dental molding gel may sometimes be substituted for the flexible silicone rubber. Hard-setting materials, such as casting polyurethane, wax and some cements may be used instead of plaster. Fiberglassed resin is used for big molds which require strength and light weight. Hot-melt glue is cheaper than silicone, but less flexible. Dental gel is both cheap and nontoxic, but may be incompatible with some casting resins. As usual, experimenting before the project will help in selecting materials.

The following table lists the most important differences between silicone rubber and plaster for molding.

SILICONE (RTV) RUBBER	PLASTER
Expensive	Inexpensive
Fast, once set	Slow—must be dried and sealed once set
Precise	Precise
Flexible—allows undercuts and multiple casts	Hard—needs positive angles for release
Needs reusable mold box	Temporary box OK
Catalyst needs precise dosing	Easy to mix
Slow cure (2 to 20 hours)	Fast (5 minutes)
Not always compatible with some casting materials	No problem once sealed
No release agent needed for nonporous materials	Release agent needed for model and cast

PLASTER MOLDING

Plaster molding is limited by the shape of the model to be reproduced. The following descriptions of techniques assume that a simple plaster female mold, in two parts, will reproduce the master model. The term "female" is used for hollow, negative shapes in design modelmaking.

Of course, there are cases when a hard mold must be made in many parts to be removable from the model (and from the cast, unless breaking the mold at that stage is acceptable). This is the case for many sculptures. Plaster dries hard, unlike silicone rubber, and planning the process must take into account the removal of the mold from both the master model and the final cast. Figure 11-2 shows section views of two plaster molds. One is possible, and the other impossible, to remove from the model.

As these simplified views show, plaster can be used for a two-part mold only for model shapes that have positive draft angles and no undercuts, which would prevent the removal of the mold from the model (and the cast). Even zero angles ("straight" sides) will cause problems with plaster molds, unless more parts are planned to allow for easy removal. It is better to design these extra parts first than to be forced to produce them with a hammer later.

To plan both parts of the mold begin by tracing a line around the model to distinguish between the two. Then press soft modeling clay around one part, and scrape off, if necessary, from the other side of the parting line, as shown in Figure 11-3.

There are many ways to proceed for plaster molding. One way is to prepare a bottomless box, built large enough to contain the model, with a thick enough layer of plaster around it for strength, but not so big as to waste plaster and end up with a heavy mold.

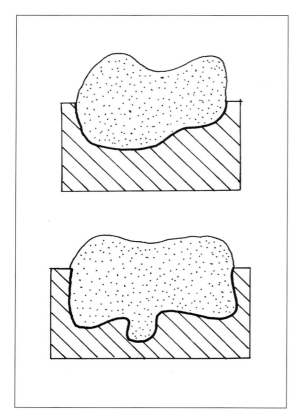

Fig. 11-2. Two-part plaster molds.

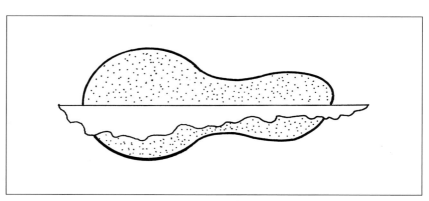

Fig. 11-3. Preparing the parting line for molding.

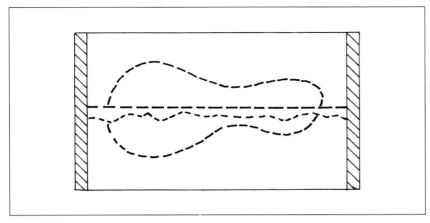

Fig. 11-4. A simple model in a mold box.

The clay on the lower part of the model should bridge over to meet the edges of the box. Shims of sheet material can be inserted into the clay for this purpose, which prevents liquid plaster from falling into the lower part of the box while pouring the top. Before fixing the model in the box with temporary supports (wires, nails, etc.), the part to be molded must be coated with a mold release compound to prevent it from sticking to the model. Some possible mold releases are paste wax, vaseline, polyvinyl alcohol (PVA) and silicone spray wax. PVA is water-soluble, making the cleaning of the finished cast easier, but it is poisonous and must be used with care. If the master model is made of a porous material, it must be sealed and dried before applying the release agent. Shellac, or another filler, may be used for sealing.

The plaster used for molding must be plaster of paris or molding plaster. Building, or masonry, plaster contains sand, so is too rough and sets too slowly for molding. To mix plaster, the powder must be added to the water and not vice versa. Sift a handful of the powder through your fingers and allow it to sink into the water. Continue doing this until the mound of plaster powder no longer sinks, then add one more handful.

After the correct amount of plaster has been added to the water, mix it by hand, kneading it slowly through your fingers to break the dry lumps. Stop mixing as soon as no more lumps can be felt. When mixing, take care not to churn air bubbles into the liquid, as too much mixing will introduce bubbles that will later appear on the mold (and cast) surface. Long mixing may also cause the liquid plaster to set too soon—it usually takes less than five minutes. Once plaster starts to set, there is no use adding more water to try to save it. The setting is a fast chemical reaction, and adding more water or mixing more will only "kill" the mix, resulting in a slurry that will never set dry, hard and strong.

A correct plaster-water mix should feel something like heavy cream, but not whipped cream, and this should be felt while mixing by hand. If it feels wrong, more water or plaster may be added while mixing.

Working fairly quickly, apply a thin coat of plaster to the top, prepared part of the model. A small amount of thicker plaster could be prepared separately for this (just put a little of the mixed plaster in a small container and mix in more plaster powder), since the creamy plaster may flow too easily off the model's surface. This first coat can be "flicked" on with the fingers, but that makes a real mess. It may also be brushed on with a soft brush. The trick is to apply it without trapping air bubbles on the surface, which will later form tiny bumps on the surface of the cast. When the first layer is on, the rest of the plaster can be poured over it to fill up the top part of the box. At this point it doesn't matter if bubbles are introduced because they will be away from the surface. The setting reaction will give off heat, and in a few minutes the plaster will be hard enough to mold the second part of the model.

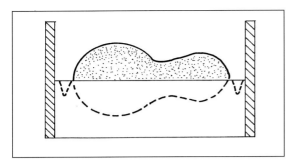

Fig. 11-5. Keys in the first mold part.

Remove the clay and shims from the lower part, cleaning it if necessary, and tip over the box so that it rests on its new plaster bottom. Then drill or gouge several "keys" with open angles into the set plaster. These keys are important, since they provide for the correct positioning of the two mold parts for the final casting.

Apply more release agent both to the model and to the exposed hard plaster, to allow for easy separation later. Put a thick layer, or several of them, on the surface of the plaster, since it is porous and may stick to the second half of the mold. The plaster should be mixed and applied in the same way as before. When this plaster is hard (when it is still warm, since heat softens the release agent for easier removal), both parts of the mold can be lifted off the model.

Cleaning up during and after plaster work must be done with care. Brushes and tools should be cleaned before the plaster sets too hard to save them. Liquid plaster should never be poured into sinks, since it will set hard in the drain pipes and require an expensive repair job. Pieces of hard plaster left on tables and floors may get walked on, and in a short time can whiten the floors of an entire buiding. Hands coated with liquid plaster should be rinsed in a lot of flowing water or the plaster should be allowed to dry on the hands, then rubbed off, before rinsing in a sink. Working outdoors with this safe, but messy material is advisable.

The finished mold must be completely dried of all the water in it that did not react with the plaster. This may be done by slow air-drying, or by drying in a warm oven. Heating too quickly or at too high a temperature (212°F, or 100°C, water's boiling point, is much too hot) can cause the hard plaster to crumble or even explode as the evaporating water tries to escape. In such an oven, drying may take about a day. Air-dried, it may take a week or more. Once dry, the pores of the plaster must be sealed with shellac or a filler, and once this is dry (two coats, usually), more release agent is added to the two mold parts for the final casting.

A pouring hole for the casting is then cut into the plaster mold. This hole must be long enough (and this comes from planning the mold thickness) to allow excess casting liquid to be poured in, since this liquid may shrink down as it hardens. The hole must also be made wide enough for easy pouring of the casting liquid without trapping air bubbles as it is poured. Later, the cast reproduction of this hole can be cut and sanded off with the flash (the feather edge along the parting line) on the cast.

It is important to plan the location of the pouring hole and the position of the mold for the casting, so that no air will be trapped under "ledges" when the liquid fills up the mold. With some designs, it is impossible to locate this hole in such a way that eliminates all ledges for pouring. Air release holes drilled downwards into these ledges will help prevent trapping air.

Finally, before casting, the two parts of the mold are secured together with tape, clamps, or by replacing the mold box around the mold.

Again, the above describes only one simple technique among many others used for hard molding. It shows the basic preparations and the precautions that should be taken before casting. With imagination and common sense, and as always, experimenting with new techniques, these basics can easily be developed for more complex models. There is also a great deal of industrial, fine arts and crafts literature available on the many molding processes used today.

SILICONE RUBBER MOLDING

For many model shapes silicone rubber molding is much easier than plaster because it allows for undercuts and negative draft angles in the design. Details are also easier to manage with silicone molds.

The rubber can often be molded in one piece around the master model through one pour. When the rubber sets, it is cut through to the model with a sharp cutter, far enough down to allow the model to be pried out of the elastic mold. The parting line for this cut should be planned beforehand for an easier release.

Silicone molding requires a mold box. It must have a bottom and should be built with screws, staples or nails for easy dismantling, rebuilding and later casting. The box should be built as small as possible, since silicone is expensive. The thinnest parts of the mold should be about $1/3$" (one centimeter) thick. The box must be rebuilt around the mold for casting because the silicone rubber is flexible enough to bulge

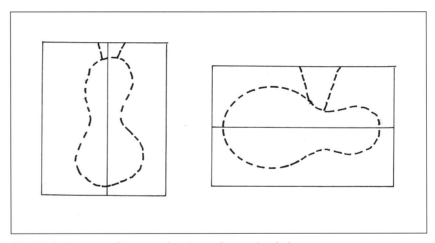

Fig. 11-6. Correct and incorrect locations of a pouring hole.

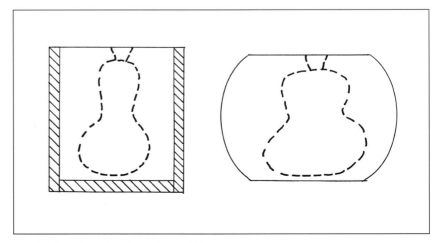

Fig. 11-7. With and without the mold box.

and stretch out of shape, due to the weight of the casting liquid poured in. The box is needed to keep the cast the same shape as the master model.

Some brands of silicone rubber work well and are easy to use. Some are worthless, even though they cost as much as the others. The worst require almost a laboratory precision to dose the catalyst, with the result that most mixes never set. The best offer a range of catalyst doses that may be used for faster or slower setting.

Mixing the catalyst with the liquid rubber should be done slowly and carefully to prevent air bubbles from being trapped in the mix. Under-surface mixers (also used for resin mixing) may be purchased for this purpose. Once mixed, most air bubbles can be removed by putting the mix in a vacuum chamber for about one minute. If left too long, chemical vapors needed for the setting reaction to take place will be sucked out. Bubbles can also be vibrated out by holding the container against a vibrating sander. Again, do not vibrate for too long, as more than about 30 seconds may cause fast hardening which will trap the bubbles.

Pouring the liquid silicone into the mold box around the model must be done slowly, so that air can be pushed up as the liquid rises. Again, planning the pouring hole location and the mold position for pouring should prevent air from being trapped under ledges as the liquid fills up. Wire or thin nails may be used to hold the model in position inside the mold box.

If it is not possible to position the model to prevent air from being trapped under all the ledges, moving and tilting the box while pouring will help free the air. The box full of liquid rubber can be vibrated again after pouring the mold to let more bubbles rise. Sometimes the pouring itself is done near a heat source, such as a light bulb or a radiant heater. This reduces the viscosity of the stream of liquid being poured, making it easier for air bubbles to escape. Pour only into the liquid—and not over the model—to give air every chance to escape.

If the model is nonporous, no mold release is needed. Nothing sticks to silicone rubber, not even itself. For this reason, mix enough to be poured at one time. The displacement methods, using fine sand or water, easily show how much volume will be needed for the mix. For example, put the model in the box. Pour in sand to fill the box, then pour that sand into a container. That's how much silicone you'll need to mix, but add a little bit more to be sure. Then clean the sand off the model. This should be done before applying a coat of release agent, when that is needed. The same displacement methods are used when preparing resin or other casting liquids.

When the silicone rubber has set, which may be from 2 to 15 hours, just enough rubber is cut down to the model to allow it to be pried out of the mold. Once free, a pouring hole is cut out of the mold, again planning all these cuts to avoid trapping air when pouring the cast. If bottom-fed injection is to be used for the casting, for example with a large veterinary syringe, air release vents must be cut out at the top and above any ledges.

Finally, a few words of caution about silicone rubber. First, some types are not compatible with chemicals such as sulfur, used in certain styling clays (you'll smell it if it's there). If a master model is made from such a clay, the rubber mold may never harden around it. Second, even a few molecules of silicone in the air can ruin paint jobs, so make sure work areas and tools are completely clean after use to prevent problems with painting.

CASTING

TRANSPARENT POLYESTER RESIN CASTING

Beginning with one of the most difficult model casting operations—that of making transparent casts—the following information will also explain liquid casting in general. A little theory, in the form of some painless chemistry and physics, should help avoid problems and make crystal-clear resin casts. But first a few words on safety with chemicals.

SAFETY NOTES

The following notes present the dangers of working with the chemicals commonly used in casting models today. Read them and follow them for your own and others' protection and insist that everyone else does the same.

- Buy and use only pre-accelerated polyester resins, which require only a catalyst to be added. Straight resin, to which both a catalyst and an accelerator must be added, can be dangerous. An explosion could result from mixing the catalyst directly with the accelerator. They must each be mixed, one by one, into the resin only.
- Transparent and opaque resin colors are poisonous.
- The resin monomer and the catalyst are both poisonous if swallowed. They may also cause skin burns and even blindness if splashed or rubbed in the eyes.
- Work outdoors or in good ventilation.
- Wear rubber gloves and wash your hands.
- Do not eat, drink or smoke while working.
- Wear a good vapor (spray) mask and safety glasses.
- PVA is poisonous, and should not be sprayed indoors. Other release agents, such as vaseline and paste wax, are safe.

Fig. 11-8. Clear cast resin and lead perfume bottles. Photo, design and models: Mona Karim.

Fig. 11-9. Sandblasted clear resin perfume bottle. Photo, design and model: Reto Christiani.

- Clean up poisonous spills to protect others.
- Avoid flames and sparks which may cause fires with any resin or solvent work.
- If a resin has been overdosed with catalyst it may begin to burn and smoke from the heat it produces. The fumes are poisonous. Carry such a cast outdoors where nobody will inhale these fumes.
- Wear a good dust mask when sanding polyester resin.
- If any of these materials get in your eyes, wash them with water for 5 minutes and go see a doctor.
- Use your kitchen oven for food only.

A LITTLE THEORY

Liquid polyester resin monomer will undergo a chemical reaction to become a solid polymer by an internal cross-linking of its molecules. In the presence of a catalyst, this reaction will go faster. The more catalyst present, the faster the reaction. This reaction is "exothermic," that is, it creates heat as it takes place (some reactions require heat instead). This reaction also goes faster if heat is supplied. As such a reaction takes place, some of this heat, if it can't escape to the surroundings, makes the reaction go even faster. The faster it goes, the more heat created, supplied, and so on. This process can be controlled to help get quality results, but it can also get out of control, resulting in a "runaway" reaction that could not only ruin a project, but could explode or cause a fire with toxic fumes.

    Repeating the above, briefly:

    More catalyst = a faster reaction.

    A faster reaction = more heat.

    More heat = an even faster reaction.

This should give an idea why some casts are rushed smoking to the nearest container of water and thrown in.

In thin resin lay-up operations, where the resin layer is only 3 or 4 mm. thick, the reaction heat is easily given off into the air or the surrounding mold. For thicker work, like casts, care must be taken to control the reaction speed.

Resin also shrinks as it hardens. The slower the reaction, the more it shrinks. This shrinkage may be as much as 8 % of the volume of the cast. Therefore, a minimum speed of reaction is desirable to avoid too much shrinking.

In a thick cast, such as a model of a bottle, not all the heat can escape quickly. In the interior of the volume, the heat doesn't escape into the mold as can the surface heat. This interior heat just makes the resin inside react faster, and the inside sets hard faster than the outside. This causes uneven shrinkage. The interior, setting fast, shrinks less than the outside, setting slowly because it loses some of its heat. It is like drying a block of green wood too fast: the outside shrinks more than the inside, causing splits and checks from the bulging stress inside. The greater the volume of the cast, the more this problem of trapped and recycled heat can cause ruined designs and runaway reactions.

Air bubbles can also ruin a transparent cast. If the resin is viscous to start with, or if it hardens too fast, bubbles won't have a chance to float up into the pouring hole, and the final cast may look similar to warm beer.

A final problem, related to the above, is that the outside layer of the cast, hardening slower because it is cooler, stays liquid longer and can stick to the mold surface. As the interior hardens and shrinks, it can pull the sticky exterior away from the mold in strands. When the mold is removed, the cast may look more like a porcupine or a sea urchin than the original design.

Fig. 11-10. Men's toilet accessories in resin and hardware parts. Photo, design and models: Marc-Antoine de Murault.

There are easy ways to eliminate all these potential problems, as long as the causes are kept in mind and controlled. Some of these controls are:

- Less catalyst can be mixed, to slow the reaction and prevent a run-away at the interior of the cast. This gives a "slow cast."
- Exterior heat can be supplied, to compensate for the reduced catalyst, allowing a safe, fairly fast hardening for less shrinkage. This outside heat will also equalize the hardening speed of the interior and exterior of the cast. The poured cast can be placed in an oven for this purpose (again, a strong light bulb in an old refrigerator will do).
- A low-viscosity resin can be used to reduce trapped air bubbles.
- The resin can be warmed before adding the catalyst to reduce its viscosity. Too much heating, though, will speed up hardening and defeat the purpose.
- The catalyzed mix can be vacuumed or vibrated to float the bubbles introduced while mixing. Too much vibrating will set the resin faster, trapping the bubbles you're trying to get rid of; too much vacuuming will pull out necessary chemicals from the resin. About 30 seconds is enough time for either operation.
- The final poured cast can be vacuumed or vibrated to get rid of bubbles poured in (putting a liquid in a vacuum chamber causes the bubbles to swell, like balloons at high altitudes, and they float up more easily).
- The poured cast can be put into a pressure chamber while hardening, to squeeze the bubbles down to so small a size that they are almost invisible.

Fig. 11-11. Clear cast bottle. Photo, design and model: Jerome Eichmann.

- The slow cast gives time to handle the operation, including bubble removal, before the cast hardens.
- The casting liquid can be injected into the mold from the bottom to reduce trapped air.
- The mold can be heated before casting to equalize the hardening speed of the cast's interior and exterior, thus making the whole cast shrink the same amount.
- Use a molding compound that will not inhibit the resin hardening (as air can do), which results in a sticky surface.
- If the resin viscosity is too high, avoid adding too much thinner, which will cause shrinkage when it evaporates. Styrene monomer is better for thinning since it becomes part of the hard polymer during the reaction.
- Before mixing resin and catalyst, transparent or opaque colors may be added to the resin. Very few drops of the color are needed. The amount of catalyst to be added will vary with the type of resin and the volume of the cast. A rough estimate for most resins, to make a slow cast of about a one liter volume, is about 0.5% catalyst. Again, the greater the volume of the cast, the less catalyst should be added.
- Pouring the cast should be done as carefully as with silicone rubber pouring, to reduce air bubbles or air trapped under ledges. Give the poured cast time to harden before trying to remove it from the mold. Sometimes this may take as much as 15 hours, even for a slow cast that has been placed in a shop oven at about 122°F (50°C).

FINISHING THE CAST SURFACE

If all goes well from the master model through molding to the final cast reproduction, there should be little or no need to finish the cast surface, except for sanding off the flash along the parting line, pouring hole and vent holes. For this work, or for any defects on the surface, sanding should start rough and work down to fine paper of 1200 to 1500 grain. Polishing follows, using several grades of polish for coarse, then fine, polishing. This can all be done by hand or with a sanding and polishing machine, which is much faster. Wet sanding will help reduce surface scratches and sanding dust.

OPAQUE CASTING MATERIALS

Many of the techniques described above may be used with opaque as well as transparent casting materials. Even the treatment of the bubble problem is almost as important, since bubbles can form on the surface of a cast. There are many materials that may be cast besides the few resins (polyester, or polyurethane, which shrinks much less) often used for casting design models. A few of these are:
- Plaster.
- Waxes (impossible to paint).
- Wood's metal (melts in hot water).
- Pottery clay (for hollow "slip casts" in plaster molds).
- Low-melting point metals (for hollow "slush casts" in cold molds).
- Chocolate and crystal candy (for temporary, edible design presentations).

Some of the non-resin casting materials have been used with very good results for centuries. They are worth considering, in view of the toxic nature of most of today's resin systems.

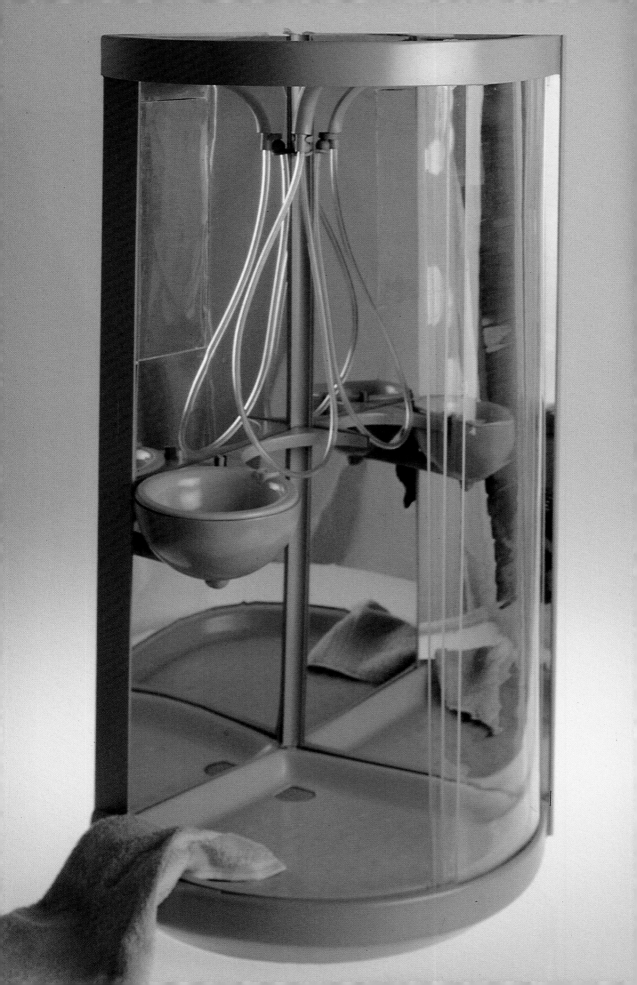

CHAPTER 12

THERMOFORMING

Photo, design and model: Marcel Soiron.

WHEN AND WHY

Thermoforming is mainly an industrial process, intended for mass or series production of identical plastic parts. An accepted part of the investment in the process, for any given product in a specific plastic, thickness, etc., is the necessary "waste" of both time and materials that goes into experimentation. It leads to an understanding and control of the specific variables in the project so that the production run will turn out consistent reproductions of the desired quality.

In design modelmaking, the aim is different. Very often, one or just a few parts must be produced. Different projects are not only more frequent, but each one is unique in its own question of how to proceed. A certain quality is desired, just as in industrial production, so some waste of time and materials is almost always necessary. The main difference is that the designer or the modelmaker does not always have the patience to see and solve whatever problems might arise for one quick job alone.

Thermoforming models, or parts of them, is only one way of reaching your objective. Often, a part may be made using another technique. In some cases, though, thermoforming is the easiest way to get good results. Some examples:

- Where hollow parts, such as bowl shapes, are needed.
- Where transparent parts are needed, and casting in resin would take too long.
- Where reproductions are required.
- Where a curved section—even a two-dimensional bend—would be too fragile or time-consuming by shaping a form.

There are many possibilities in the size and shape of thermoformed parts. Pleasant surprises will definitely outweigh first-time frustrations in working with the process.

THE BASIC OPERATION

Thermoforming means heating a thermoplastic sheet (it won't work with thermosetting plastics) to its forming temperature, then forming it into a desired shape and cooling it so that it keeps that shape. Forming can be done mechanically (bending, stretching or pressing against molds), with air pressure, with a vacuum, or with a combination of any or all of these. It can be as simple as spot-heating with a hot air gun or using boiling water for PVC. Most often, a vacuum forming machine is used because it can not only heat large sheets uniformly, but can also form the sheets into shape against molds. The machine need not be automatic. A reasonably good "manual" machine should have:

- Heating elements to soften the plastic sheet from above and below.
- An adjustable clamping frame to hold plastic sheets of various sizes, which can be slid out from between the heaters for the forming.
- Platens connected to air vacuum and pressure, which can be raised or lowered to meet the softened plastic sheet.
- A vacuum pump with a good-sized "surge" tank.
- A source of compressed air.
- Safety devices to protect the operator.

Usually, a mold must be made to form the plastic part, or a shape can be made by blowing a bubble. This may be male or female, as will be discussed further below. It will also be shown that there are certain limits in the material used for the mold, as well as the shapes possible. These are design limitations inherent to thermoforming. A mold box is needed to hold the mold to the platen to allow vacuum or air pressure to do the forming. A sealing frame is also needed to create an efficient vacuum.

THE PROCESS

A sheet of plastic is clamped tightly in the frame and slid inbetween the heaters. Figure 12-2 shows the plastic sagging down upon softening. It is

Fig. 12-1. A vacuum forming machine.

better to train the eye to watch for the correct sag than to try to use timers and timing tables. Of course, thick plastic takes longer to heat uniformly, and some plastics' high forming temperatures require longer heating. The process described here deals with manually operated machines used for simple models of one (or limited) reproduction. Continuous-sheet machines, and even timer cycle controls on even some simple machines are not needed for model work. What is needed is constant attention and the ability to learn and correct through experimentation, since almost every model will be very different from the last.

Once the sheet has sagged down to about one quarter to one third of its cross section, pull the frame out from the heaters to form the part. It is important to constantly watch the sheet being heated. If it sags too far down it may touch the bottom heaters, ruining them and maybe even causing a fire. Experience will also show that when various plastics (e.g., acrylics and polycarbonates) are being heated too fast, bubbles form in the sheet, in which case the temperature controls of the heaters must be adjusted for slower heating. When the softened sheet is pulled out, the part is ready to form—immediately— before the sheet starts to cool.

There are many possible variations of the forming process. A few—the most useful for modelmaking—will be described here and the variables which affect the work will be presented later. For the complete range of thermoforming possibilities, which grow every year, industrial literature may be consulted. Briefly, a few of the variations are:

Fig. 12-2. Softening a plastic sheet.

- The mold may be attached to either the top or bottom platen (for most work it is on the bottom).
- The mold may be male or female.
- Either air pressure, vacuum or both may be used to press the soft plastic against the mold.
- For deep shapes and some types of detail, a combination of using both the top and bottom platens may be used.

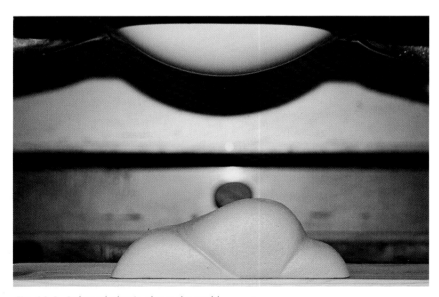

Fig. 12-3. Softened plastic above the mold.

Fig. 12-5. Vacuum pulls the plastic against the mold.

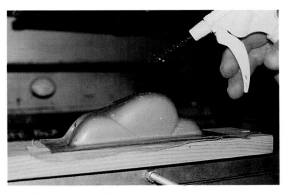

Fig. 12-6. Cooling, then releasing the part.

Fig. 12-4. Platen raised to push mold into the plastic.

Drape forming, the simplest of all thermoforming techniques, can form many different types of parts. Figures 12-3 through 12-6 show the stages of the process. With this basic technique, the male or female mold may also be attached to the top platen. This allows the air seal frame to be placed on the bottom platen, and once the mold is lowered onto the sheet and against the seal, no manual operation of the sealing frame is needed.

Once the cold part is released, it must be trimmed to remove the unwanted plastic (called "flash") around the edge. Acrylics, transparent PVC and styrene are cut with fine-toothed, high-speed power tools. PVC foam, opaque styrene and acetate can be trimmed simply by scoring and breaking off pieces.

There are other techniques that are used to take advantage of the properties of softened plastic. Any thermoplastic sheet, if it is heated uniformly to its forming temperature, can be stretched evenly, keeping an even thickness throughout. If parts of the sheet are heated less, or if they touch a cool mold, they will stretch less and remain thick, creating thinner areas where hot plastic continues to stretch toward deeper parts of the mold. This may be both a limit or an advantage. One simple example is where a plastic sheet is pulled into a female mold to form a bowl shape, as shown in Figure 12-7.

As the hot plastic is pulled by the vacuum into the female mold, it first touches the rim, where it cools and remains thick, resulting in a strong rim. As the hot sheet continues to stretch toward the bottom of the mold, it becomes thinner, resulting in a thin, weak bowl bottom. In some cases, especially if the mold is deep, the plastic at the bottom will be stretched too thin for strength, and may even pull apart during the forming.

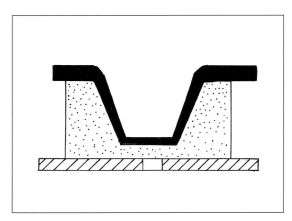

Fig. 12-7. Forming a bowl in a female mold.

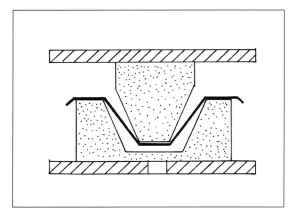

Fig. 12-8. Forming with a plug.

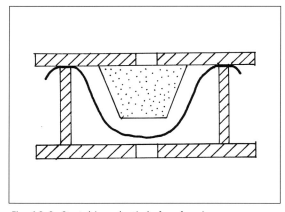

Fig. 12-9. Stretching plastic before forming.

One technique used to compensate for uneven stretching is forming with a "plug." A female mold is on the bottom platen, and a plug (a smooth male mold, smaller than the part to be formed) is attached to the top platen. When the female mold is raised into the plastic sheet and sealed with an air-seal frame, the plug is lowered, pushing the plastic toward the bottom of the female mold. A vacuum is then quickly pulled into the female mold. The result is thicker plastic at the bottom of the bowl, since the cool plug has reduced its stretch.

Another technique can be used to allow uniform stretching of the heated sheet before it touches any part of a mold. As Figure 12-9 shows, the hot plastic is first pulled into a deep air-seal frame on the bottom platen. When the mold on the top platen is lowered into this evenly-stretched bubble, and sealed against the lower air seal, the vacuum is then released on the bottom (and replaced, if necessary, by air pressure), and a vacuum is pulled at the top, pulling the bubble into the mold. This allows for deeper shapes than with drape forming.

Variations of this technique can also be used. One is the "snap back" forming method. Instead of a vacuum pulled to lower a bubble into the bottom air seal, air pressure is used to blow the bubble up toward the mold at the top. The mold is then lowered down into the soft plastic and against the seal, and the vacuum is pulled at the top. Because of the cooling action of the mold, and the reduced possibility of the plastic folding as it changes direction toward the mold, deeper shapes can be made using this technique.

In a few cases, a "free blow" method— without a mold—may be used to form a bubble shape. Skylight domes in buildings and aircraft canopies have been formed this way.

Mechanical forming can be done with the machine using only the heaters. Matching male and female molds are made to press the plastic into shape without using vacuum or air pressure. Some simple, two-dimensional curves are best done this way. Care must be taken to mis-match the two molds by the thickness of the plastic used, and to allow for longer cooling if the molds are made of foam or another insulating material. If such molds are removed too soon, the plastic may warp out of shape when it does cool completely.

Again, the above techniques are only a few of the many used in industrial thermoforming today. In almost all cases, these few are sufficient for modelmaking. Other techniques include using preformed plastic tubes instead of sheet for bottle forming as well as combinations of the methods described above.

CONTROLLING THE VARIABLES

TYPES OF PLASTICS

Although all thermoplastic sheet may be formed with heat, some types are easier to work because of their lower forming temperatures. Some are difficult for other reasons. For example, water in or on the plastic may cause bubble marks when heated. The following are a few of the sheet plastics commonly used in modelmaking and some of their characteristics:

- Cellulose acetate: Forms at about 302°F (150°C). No real problems.
- PVC: Can be formed with boiling water if necessary. The clear type is too brittle to score and break for trimming.
- Acrylic and polycarbonate: Both form at high temperatures and resist tight bends. Both may cause bubbles if not previously dried or heated very slowly.
- Polystyrene: Low forming temperature (302°F, or 150°C). Easy to form. The opaque styrene may be trimmed by scoring and breaking.
- Foam sheet: Thin sheets, such as PVC foam, are easy to form. Must be heated slowly.

This list would be much longer for industrial applications, since there is such a wide variety of thermoplastics available today.

THE MOLD

Thermoforming molds can be made of various materials, each of which presents both limitations and advantages. In design modelmaking, the more durable molds in hardwoods or aluminum are almost never used. Plaster, urethane foam, fiberglassed polyester and cast urethane are often used. It is important that if the mold is a good conductor of heat, it must be preheated to prevent cold spots during the forming. Also very important is the heat stability of the material or its coating. Plaster molds must be completely dry to prevent heated water from forming blisters in the soft plastic that heats the mold on contact. Thermoplastic materials, such as styling clay, wax, styrene foam and even surfacing coats, such as thermoplastic fillers and primers, will all melt or even boil with the heat of the plastic. This may cause deformations or blisters in the formed part.

The placement of airholes for the vacuum is a variable that requires careful planning. Thin plastic sheet requires small holes, as large ones would create too great a vacuum force, thereby causing deformations. The spacing of the holes and their locations must take into account the shape and details of the mold. More holes are needed in deep "corners," where sealed air pockets could cause a bridging over of the plastic due to a lack of vacuum in such areas. Instead of drilling holes once the mold is made, it is sometimes easier to include greased wires in molds that are cast, for example in plaster or solid urethane. Once the mold has set hard, the wires are pulled out, thereby making the holes. Fly screen can also be placed between the mold and the mold box. This allows air to flow under the mold during the vacuum forming, and avoids having to make matching air flow holes in the mold and mold box.

Molds with smooth surfaces will produce smooth parts on the surface of the plastic that contacts the mold. Rough or matt molds, such as unfinished urethane foam, will produce rough, matt or frosted surfaces upon contact, which may be just what the design calls for. In most cases, a mold release compound is needed to allow easy separation of the cooled plastic part from the mold. Vaseline, talcum powder, silicone spray, or soft stretch cloth may all be used. Cloth also provides a smooth surface instead of finishing a rough mold. Vaseline, brushed on, then melted smooth with a hot air gun, may also save on surfacing time. The same vaseline will melt again with the hot plastic. It will help to remove the part from the mold while the vaseline is still soft.

Design limitations in thermoforming are also a part of planning and making the mold. Tight inside corners and edges do not reproduce well in formed parts. For this reason, industrial products always have at least slightly rounded edges. Contoured details involving "undercuts," or negative draft angles, are also impossible to form, unless the mold is destroyed to free the part. This is always risky, since the part may be destroyed while trying to free it, leaving no part or mold and setting the modelmaker back to square one.

Male and female molds may be made for thermoforming. The male mold, positive and convex in its general shape, is easier to make. Its surface will replicate itself on the inside of the formed plastic part (smooth, rough, etc.), which may or may not be desired. Since plastic shrinks on cooling, a male mold needs to be designed with a generous draft angle to allow release of the cool part after forming. A female mold, hollow and negative in general (the unfortunate terminology was created by the industry), is more difficult to shape and finish. Its surface will give a matching finish on the exterior of the plastic part formed, which again, may or may not be desired. For example, if a bowl shape is to be produced having a visible, convex exterior smooth finish, a female mold having the same finish must be used. In such a case, an easier, male mold could first be made, then used as a mold to cast a female mold around it. Female molds release formed parts more easily than male molds, again because of shrinkage. This allows for designs with "straighter" sides, having fewer draft angles than needed with male molds.

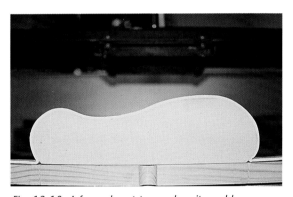

Fig. 12-10. A formed part trapped on its mold.

The depth of vacuum (or pressure) pull, and the required thickness of areas on the formed part will also determine whether a male or female mold must be used. Keep in mind the even, free stretching of hot plastic and the cooling and stretch resistance of plastic when it first touches a cool mold part when designing the mold.

TEMPERATURES AND PRESSURES

Sheet heating temperatures must be kept close to the forming temperature limits of the plastics used. If a sheet is too cold, it may tear upon stretching or may resist being pulled into details and corners. If it is too hot, it may melt or flow apart (or even burn) when heated and stretched. The entire sheet must also have a uniform temperature, even if this means preheating a mold or the clamping frame to eliminate cold areas. When air pressure is used instead of vacuum to form a part, the cold air might cause uneven stretching.

Air pressure, or vacuum, must be great enough, with enough volume (from a big compressed air or vacuum surge tank) to generate a fast push or pull, even with large models. Slow forming may let the plastic cool before the work is done, resisting shaping into details and far reaches of the mold.

In the mold, and in the vacuum box if screen is not used, the holes should be big enough to allow a quick pull, but not so big as to deform the thin sheets. Again, more detail and inside corners require more holes spaced closer together. If soft coth is used on a mold surface, or if the mold is porous, there may be no need for any holes.

The remainder of the system—mold to mold box, mold box to platen, etc.—must be airtight, with no holes, for a good fast pull. Masking tape, neoprene weather strippng (not the foam type), modeling clay or vaseline may help. Particle board boxes should be sealed with a surface coating.

SHEET SIZE

The clamping frame on the thermoforming machine is usually adjustable and can be made to fit large or small sheets for various projects, a feature that saves money on plastic. Sometimes adjusting the frame takes too much time, or the sheet size is too small for the minimum adjustment. Small parts can then be heated on an aluminum sheet placed on the frame, and laid onto the mold by hand for forming. If uneven heating occurs, the small sheet can be flipped over once in a while and slid back into the heat source.

SAFETY

Like all electrical and mechanical equipment, the thermoforming machine must be used with care. The following are the most common sources of danger along with precautions that should be taken:

- Hot surfaces: The heaters, the clamping frame and the heated plastic may be very hot. Wear gloves and never reach into the heater part of the machine.
- Moving parts: The hydraulic presses which move the platens up and down are powerful enough to cause severe injury. Keep hands and arms out of reach. Someone else may actuate a platen when you least expect it.
- Water: The heating elements are often very powerful. Water, used to cool parts fast, should only be sprayed lightly or applied to the plastic with a slightly wet sponge to avoid splashes that may cause steam scalds if water falls on the heaters. If there is a fire sprinkler system installed overhead, insist on having it removed or refuse to work with the machine. The heaters can set off the sprinklers, resulting in the danger of electrocution or steam burns.
- Volatile solvents: Their presence in the area can cause an explosion or fire because of the high temperature of the heaters.

COMMON PROBLEMS

Were we to compile a complete list of potential problems, causes and solutions in thermoforming, it would be very long indeed. The following are the most common problems that you might encounter when forming design models:

- Mold and part stick together: Design rules ignored (for draft angles and details). Release agent forgotten. Thermoplastic mold or its finish melted against sheet. Part removed too late.

- Poor detail: Vacuum too slow. Air leaks. Sheet too cold. Details too fine. Mold too cold. Airholes too small, too far apart or located in the wrong place. Sheet too thick. Wrong choice of plastic. Wrong choice of male or female mold.
- Areas too thick or too thin: Cold spots by contact with mold. Uneven heating. Shape too deep for technique chosen.
- Surface marks: Mold not finished enough. Wrong choice of male or female mold. Dust in the area, on the mold or on the sheet being formed.
- Blisters and deformations: Thermoplastic mold or surface. Wet or damp plaster, etc., mold water boiled. Water in the compressed air.
- Bubbles in transparent sheet: Acrylic or polycarbonate sheet not dried long enough or heated too quickly.
- Sheet melts or scorches in heaters: Sheet too thin. Temperature set too high for the plastic used.
- Clear sheet whitens: Cold spots being stretched. Temperature too low. Pull too slow.
- Folds or wrinkles: Wrong technique for mold shape. Temperature too high. Vacuum too low. Air leaks.
- Removed part distorted: Part not cooled enough before removal. Mold too warm or too good an insulator. Two molds (for mechanical forming) not left on long enough.

Again, there are sure to be other complications with some of the designs you will attempt. Be prepared to lose a few sheets and learn from your experience.

A FEW EXAMPLES

Figures 12-11 and 12-12 show thermoforming projects made with several materials and techniques. One was the result of careful planning and experimenting, and the other was just plain good luck.

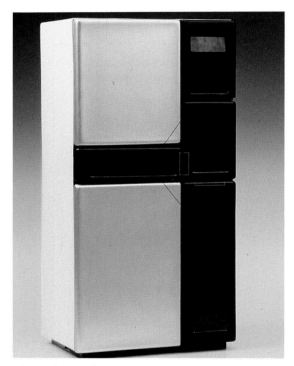

Fig. 12-11. A refrigerator model made by thermoforming. Photo, design and model: Philippe Holland.

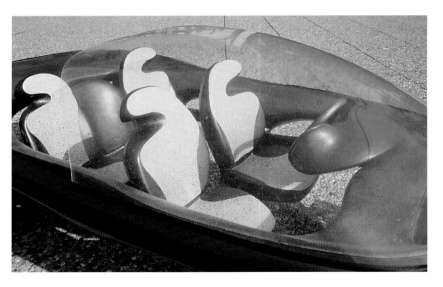

Fig. 12-12. A very lucky day: the mold was made of thermoplastic clay. Photo, design and model: James Dean.

Figs. 12-13 and 12-14.
A model made with
thermoformed parts.
Photo, design and model:
Stephan Messmer.

Drawing by William Wright.

CHAPTER 13

MAKING

SWEEPS

HISTORY

Like the classical French curve used by draftsmen, sweeps are used by designers for tracing forms and curves. Usually made of transparent, flexible plastic, they are useful for drawing because you can see through them as you work. Automobile sweeps were originally patterned after the curves of ships. Today, automobile sweeps are different. The usual set of nine or ten auto sweeps (there are a few others, some very large for full-scale design) gradually developed independently from nautical sweeps but stopped evolving, the result being that the same design lines have been reused over and over again. This is understandable for ships and boats, where hydrodynamics always played a necessary part in design. In the auto industry, though, this consistency is harder to appreciate, especially today with all the "new" car designs which are sometimes hard to tell apart, especially from the rear. One European auto company president recently said that such homogeneity is mainly due to government restrictions on heights, widths, light signal positions and sizes. Maybe so, but another specialist—a well-known car designer— also said that you can look through the history of car design, placing your sweeps against photos in styling magazines, and see that the same sweeps have been used for years. In any case, using sweeps to draw curves is a far cry from carrying a set of radius gauges around in your pocket for all your designs. That just amounts to a blind obedience to industry, in its well-known lack of imagination and insistence on maximum production profits.

But this chapter is intended to show how to make sweeps as well as judge their merits. A good set of sweeps can be very expensive, and in many countries may be very difficult to find. Another reason for making sweeps is that the work offers valuable practice for other modelmaking work. It develops patience, as well as special edge-sanding and checking techniques. It is also a good example of deciding that second-best, or "almost good enough," is not really good enough for the purpose. Good sweeps have to be really good, because they may be used for many years.

TRACING, SCORING AND BREAKING

To begin with, it is necessary to borrow a set of the best sweeps you can find, promising that no cutter will come near them (a cutter can scratch and damage a sweep, throwing off its curve). Using these originals, trace each sweep with a fine pen that will mark a sheet of 1.5 to 2 mm. flexible plastic (do not try marking it with a cutter). Butyrate or cellulose acetate are best because these plastics can be lightly scored along the traced ink line with a fine point or a cutter, then broken apart to produce the basic outline. Acrylic, transparent PVC or transparent styrene will not work because they shatter when broken. Instead, these need to be cut with a fine-toothed band saw or jigsaw blades. The hooked scraper that many Japanese modelmakers use is also a good tool for scoring, with many repeated passes, if a brittle plastic must be used.

The traced line must be fine—so that there is no confusion as to which side of the line is the outside and which is the inside—and the cut must be made precisely outside the fine line. One pass of the cutter should be enough. Repeated passes will just bind the sharp blade's wedge in the cut, unless the sheet is bent open while cutting. Scoring and breaking are a little more difficult for tight, inside curves. Needle-nosed pliers may help to break such areas. Start breaking by bending from one end and continue along the curve as it breaks softly apart. Leave a generous space between each sweep when tracing the lines. If only a few millimeters of waste plastic separate the sweeps, breaking off this narrow strip may very well distort the sweep.

Figs. 13-1 and 13-2. Scoring a traced curve in styrene with a sweep. Once the figure is scored, the excess styrene can be broken off. Photos: Micaela Porta. Design: Alex di Nota. Reproduced by courtesy of Henry Dreyfuss Associates.

SANDING AND CHECKING

When all the sweeps are broken out of the sheet, the real work begins: sanding down to the line, correctly. If there is anything a design modelmaker learns to do—and do a lot of—it's sanding.

It is less important to precisely follow the original line than to produce a true, continuous curve. You may end up with a curve that is slightly different from the one you traced—and produce a real original—but that is far better than a sweep with flat spots that break the curve, which will render the sweep practically useless.

Sanding should be done by hand along the edge of the sweep. It may help to clamp the plastic between wood braces while sanding and checking the curve. This must be done in good light, making use of the very light reflections that need to be avoided when drawing for a long time. Light glare tires the eyes when drawing, but this glare is necessary for good sanding because it allows for the thorough checking of smooth surfaces. Stop sanding often, wipe the sweep clean and look at it lengthwise—like aiming a rifle—into the light. Keep checking to see if there are any flat spots or sudden changes in direction, which must be eliminated. Begin sanding with rough paper, gradually moving toward about no. 180, using a sanding support behind the paper bridge to remove imperfections. For the final fine sanding, finger pressure can be used instead of a sanding support. This will produce a slightly rounded edge, which is desirable. It will also prevent inked lines from being sucked under the sweep during drawing. Final checks should also be made by using the sweep to draw very fine lines with a hard pencil and examining the results.

Fig. 13-3. A sweep's curve is checked during sanding. Photo and sweep: William Wright.

Again, the continuity of the curve is most important, so that the entire sweep can be used. One that has only been rough sanded, or one that has been sanded across its edge, will be too jagged to draw smooth lines. If you have sensitive fingertips, another good way to check is to run your fingers along the edge of the sweep, feeling for flat spots and sudden curve changes.

Once the sweeps are finished—and this is not a quick job—they must be cared for properly. If a few scratches on the surface distract you, sand the entire surface matt with very fine paper. This will not make the sweep opaque for your drawing work. Keep the sweeps away from sources of heat like hot sun, closed car trunks and radiators, which may cause the thermoplastic to deform. If they are good sweeps, you can contribute to design history by lending them to others for tracing.

Fig. 13-4. A set of finished sweeps commonly used in transportation design. Photo and sweeps: William Wright.

Fig. 13-5. Drawing made using several sections of different sweeps. Photo, design and drawing: William Wright.

Fig. 13-6. Drawing made using several sections of different sweeps. Photo, design and drawing: William Wright.

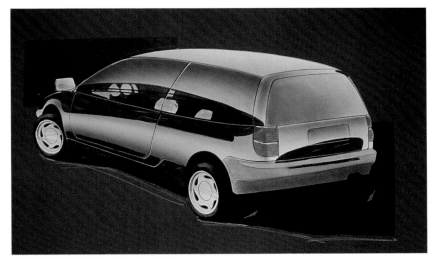

Fig. 13-7. Drawing made using several sections of different sweeps. Photo, design and drawing: Shu Takahama.

Photo: Micaela Porta. Reproduced by courtesy of Henry Dreyfuss Associates.

CHAPTER 14

MACHINES AND HAND TOOLS

MACHINE SAFETY

Fig. 14-1. Dressed for machine work with hair a clothes secured and no jewelry. Photo: Monique Carroll.

This chapter is extremely important for students, professionals or anyone doing shop work. The information presented below in no way constitutes an apprenticeship or instruction course in tool or machine practice. Good, hands-on teaching is imperative. Learning and working with experienced instructors will not only benefit you, but also those with whom you work. A good rule of thumb for anyone, including teachers and managers, is "no rushed work." The following sections deal only with safety and the prevention of accidents.

Perhaps the best way to introduce the subject of safe machine work is to advise you to be alert at all times when using a machine. It is preferable to be wary for a period of time than to approach machines with little or no care. Undue confidence, lack of sleep, a little alcohol or drugs, and even uncaring managers or teachers with selfish rush deadlines are all good reasons for simply deciding not to use machines. So are loose clothing, jewelry and hair, all of which can get caught in moving machine parts. Learn to memorize the STOP switches on each machine in case of an accident. Once you've finished using a machine, switch it off and stay by it until it stops. Machine work can save a lot of time and produce some very good results in design modelmaking, but it must be done correctly, always and without exception.

You may be surprised to find that many people have learned different ways of working with machines, or that they haven't really learned much at all. The accepted norms differ from occupations and countries. Share and practice the following information with friends and colleagues, for your own and their protection.

THE CIRCULAR SAW

Portable circular saws are not only practically useless for most model work but they are also quite dangerous. You would fare better using a table saw, remembering that this machine, too, is dangerous and that precautions do need to be taken. Wear safety glasses with plastic lenses (even over another pair of glasses), never wear gloves and stay away from the turning blade by more than an arm's reach. Use a pusher stick or a clamp to hold the piece being cut; should it suddenly split open, the pressure you're placing on it could propel your hand toward the blade. Splitter and blade guards (covers) are available and should be used, even though they are somewhat uncommon in the United States.

Keep the blade low—just half an inch above the table—and ask for help when ripping long work in order to keep it flat on the table during and after the cut. Never (never!) reach for a cut-off piece of material; flick scraps off the table with a pusher stick instead, or just leave them where they are. The miter guide or the carriage should be used for all cross-cuts, pushing on that side of the blade only. Severe kickback could result from pushing work with both hands, applying pressure on both sides of the blade. When using the rip fence as a length stop (guide) for repeated cross-cuts, be sure to install the parallel fence well forward of the blade to prevent cut-off pieces from being pinched between the rip fence and the turning blade. This could break the blade, or at least send the cut-off piece hurtling toward you at high speed. Use only very sharp blades to avoid trying to force work to get it cut. A good sharp blade makes a buzzing sound as it cuts and offers little resistance to feeding the work into the blade. A dull blade screams and smokes as it tries to burn its way through the work. When cutting very small pieces, use a smaller blade and reduce the slot in the table, using a clamped piece of

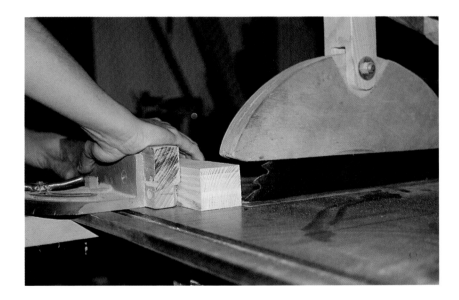

Fig. 14-2. A crosscut on the circular saw, using the miter guide and the protective blade guard. Photo: Monique Carroll.

Fig. 14-3. Doing everything the wrong way: No guard over the grinding wheel; no tool support; and holding the part by hand instead of with a clamp or pliers. Photo: Monique Carroll.

plywood or plastic sheet if necessary. Don't use a broken or split blade, even if someone has tried to repair it by drilling a hole at the end of the split, and use the blade guard as a hold-down to prevent work from lifting up while cutting. Different materials require different blades, so make sure you're using the correct one for the job. A blade for acrylic plastic, for example, has many small teeth to prevent shattering the material during the cut.

THE SHAPER

The wood shaper, using high-speed blades to cut profiles lengthwise (or worse, against the grain across or on contours) has no place in the model shop. The main cause of missing fingers, the shaper cuts at a very high speed close to your fingers, instantly severing them should the tool's blade shock and shatter the wood, allowing your hand to rush toward the cutters.

THE TABLE, OR BELOW-SURFACE, MILLING MACHINE

This is another machine that can be dispensed with. It is extremely dangerous and not worth the risk for the little it can offer. One pushes the material by hand toward a very high-speed cutting tool in order to contour profiles in wood. If you use this machine, make sure the material is securely clamped down and that you move the mobile machine along the material, not vice versa.

THE SURFACE PLANER, OR JOINER

The minimum length of stock, or board, you should try to plane is about 19^1/$_2$" (50 cm.). Shorter pieces may lift or rock down into the blades and cause a very serious kickback. Don't plane hard, brittle materials, such as acrylics, cast urethane or clear PVC and styrene because they may ruin the blade, shatter or kickback as well. Wood with knots is also dangerous, as it could break in your hand right under the blade. Thin material should be avoided because it forces your hand too close to the blade. Remember

Fig. 14-4. Planing wood flat under the blade guard. Photo: Monique Carroll.

Fig. 14-5. Asking for real trouble by trying to hold a metal piece for drilling. Photo: Monique Carroll.

to always keep your hands away from turning blades, pushing the material down and forward. Develop a rhythm, pushing the stock almost 8" (20 cm.) from the blades, then pulling with the other hand the same distance past the blades, pressing the stock down flat all the while. Use the blade guard, just above the stock for wide work, or with the stock sliding between the guard and fence for narrow work. Plane *with* the grain of wood. To do this you must examine the fiber direction, then plane (and change direction if necessary) to press the fibers against the board, rather than lift them off, which could result in dangerous breakage of the material over and behind the blades. A maximum of 2 or 3 mm. should be planed with each pass.

Don't try to acheive a uniform thickness along a board with the planer, or plane out bends or "propellor" twists (although it may help to plane only a concave side). The thickness planer will do the dimensioned work once the surface planer has flattened one or two sides. When in doubt, ask for help from a competent operator.

THE THICKNESS PLANER

This machine reduces material that has already been planed flat on one or two sides to a uniform thickness. Most machines have feeder rollers that pull the stock against the turning blades. Keep your hands away from the blades—just let the machine do the work, lifting a heavy long end slightly to keep the work flat in the machine. Maybe a wipe with kerosene or paraffin oil on the table, with the machine stopped, will help narrow stock slide better. Again, don't plane off more than 2 or 3 mm. at each pass or feed in short (less than 19¹/₂", or 50 cm.) pieces, and resist the temptation to save time by planing several pieces at once. If the machine is a combination (surface and thickness planer, etc.), be sure all other moving parts are covered with proper safety guards. If there is a chip extractor connected, use it to keep the cutting blades clear. Then turn it off to prevent deafness (it can get as loud as 90 decibels, as can the planer).

THE DRILL PRESS AND MILL

Safety glasses should be worn when milling, and keep your hands and hair away from the turning chuck and bit. Never hold a piece of work by hand for drilling or milling. As the bit cuts through the bottom side, it catches and exerts an enormous turning force which tends to spin the work, possibly resulting in serious injury. Instead, make sure work is solidly clamped or bolted to the table—in several places—before turning on the machine. If the chuck requires a key to tighten bits, be sure to remove the key before starting the machine. Don't try to advance too fast with the mill bits, and be sure the advancing stock is going in the opposite direction of the cutting edge to prevent breaking the bit as it spins.

THE LATHE

Be sure the material is at least nearly centered before turning on the machine. Keep the tool rest close to the reducing work, which means stopping the machine often to adjust the rest. Remove the tool rest when sanding to avoid pinched, sanded or broken fingers. Use low speeds for big work, especially before it is turned round. Feed the tool against the work lightly to remove only a thin film of material at each pass while holding the tool firmly in both hands, down against the tool rest. If there is a chuck key, don't forget it in the chuck before switching on the machine. If the lathe has an automatic feed (for metal or plastic work), be sure you understand the safety stops and set them to prevent overriding, which can break spinning machine parts.

When using glue for faceplate work, let the glue dry completely for strength.

SANDING MACHINES

Although it is not really dangerous to work with big belt and disc sanders, sanding yourself can be very painful and can take a long time to heal. When using a fixed stop, keep it adjusted close to the moving belt or disk to prevent fingers or small work (which should be secured with clamps or pliers) from being pulled into the space. Be sure such a stop is fastened tight. Wear tight clothes, no jewelry and keep your hair pulled back away from the machine. A good dust mask is a must, and make sure your work space is well ventilated. Glass and small metal pieces should not be sanded, and be sure that all motors (including small power equipment) being used in the area are explosion-proof. Refrain from smoking, as with any machine work.

THE BAND SAW

Don't back up work after cutting partially through it; this can pull the blade off the wheels and break it into flying pieces. Stay more than an arm's length from the blade, using a pusher stick or clamp to advance the material, and keep the blade guards perfectly adjusted, with the upper guard just above the height of the work. Blades should not be over-tensioned, keeping in mind that thin blades must be less tensioned than wide ones. When cutting small curves, don't cut them too small for the width of the blade being used. You shouldn't even attempt to cut round stock unless it is glued or fastened to a flat base so that it won't roll with the blade. Only flat-bottomed surfaces should be cut, firmly sliding on the saw table. If the work is fed too fast, it can catch and break the blade. Again, let the machine do the work. If it refuses, your blade might be dull and simply needs changing. Have the blades sharpened only by professionals, but at the correct tooth angle—a blade with teeth sharpened at too great an angle may hook into the work and possibly break. Materials such as foam won't dull the blade quickly (a rip fence should not be used for length cutting of anything harder than foam), but a blade turning for a long time around the wheels will develop tiny cracks between the teeth due to constant bending. These cracks could break the blade but can be sharpened, or ground, out. (If a lot of foam or other soft material is being cut over a period of days, remove the blade and send it out for sharpening, even if it is not dull.) A narrow blade heads where it wants to; freehand cutting along marks is best.

THE THERMOFORMING MACHINE

The chapter in this book on thermoforming describes the special safety rules that should be followed with this work.

THE GRINDER

Even for quick jobs, safety glasses should always be worn when working on the grinder. Don't try to hold a piece of metal for grinding by hand—use strong pliers or a clamp to hold the material instead. Keep the rest adjusted close and firmly attached in front of the turning wheel. If the wheel has been cracked or broken, replace it and throw it away. Also, be especially careful not to produce grinding sparks where dust or solvents in the air might explode.

Fig. 14-6. Changing a bit with the drill chuck key. The electric cord and plug are where they should be. Photo: Monique Carroll.

HAND POWER TOOLS

Small power tools are widely used. Because of their popularity, it would be wise to keep in mind a few basic rules which apply to most of these machines. Always unplug an electric tool before making any adjustments to it, such as changing the blades, and be sure that when plugging it in its switch is turned to OFF. When using a chuck key, remove it from the chuck before switching on the machine. Again, do not hold work by hand, but firmly clamp it for drilling, sawing, routing and sanding. The machine, however, should not be clamped in a vise for handheld work. Stay away from wet areas when working with these machines, and take the usual precautions of wearing safety glasses, tight clothing and hair, no jewelry and being in a comfortable, stable position (standing up) for all work. To keep them in good condition, electric cords should be protected from contact with moving machine parts.

SPARKS, SMOKING, SOLVENTS AND DUST

All power tools produce sparks from their motors, yet very few are explosion-proof. In very dusty areas, or where solvents are being used (and design modelmaking areas qualify), sparks, flames and even a lighted cigarette can cause explosions. Good ventilation, regular cleaning up and separation of work areas are as important as individual work safety procedures.

MACHINING TIPS

It has already been suggested that a good apprenticeship is needed to develop expertise in machine work. In addition to this, years of practice will give you more control over machine operations and will help you to develop the "tricks" that each operator adds to his/her personal approach to machine work. The following brief outline may serve as a head start, or a supplement, to the slow but enjoyable process of learning about machines.

MACHINES IN GENERAL
- Do not work with a machine unless you know the basic safety rules for its use. Ask questions, or ask someone to do the work for you.
- Don't use compressed air to clean machines. This will often force abrasive dust into the gears and bearings.
- Keep circular saw, planer, etc., tables lubricated for easy sliding.
- Check adjustments once in a while.
- Don't rely on initial machine graduations for precision. For example, a pointer may have to be bent to make a marked graduation meet what the real cut shows.
- Let the machine guides act as stops for repeated operations in order to avoid measuring and marking.
- Be sure to keep a flat surface or edge as a reference during all dimensioned work.
- Use only sharp cutting tools, both for safety and for quality work.
- Invent, make and set up jigs on machines for repeated work to save time and to improve precision.

THE CIRCULAR SAW
- Have two machines if possible, one small one set up for fine work.
- Use appropriate blades for all work to improve on quality and to save on work and sharpening time.
- The blade guard (cover) is not only a safety device but a help. Use it to hold work down flat on the table as you saw.
- Clamp a piece of styrene or plywood to the table to reduce the blade slot for fine work.
- When cutting acrylics, use a slow feed, a lowered blade, fine blade teeth and lubricate the blade with paraffin wax. Clean the plastic afterwards.
- When cutting cast polyurethane, treat it as you would wood since it may warp. For example, rip-cut (or plane) two long faces to equalize the warp produced by the cutting heat.
- Keep work flat against the table, getting help if necessary for big cross- or rip-cuts.
- Reduce lengths of stock for rip cuts by rough-cutting (with a sabre saw) beforehand.
- Don't assume bought stock (plywood, particle board, etc.) is square. Mark and cut from one (only) flat reference side.
- For angle cuts with the miter guide, use a hold-down clamp or glue sandpaper to the guide face to prevent slipping and rounded cuts.

THE BAND SAW

- Try to have at least one machine big enough to cut standard foam blocks (usually about a $19^1/2$", or 50 cm. depth of cut).
- Always saw with flat surfaces down against the table. Glue wedges under spaces in the stock if necessary to accomplish this.
- Use a fine-toothed blade for brittle plastics.
- Make a thin, flat rip fence extension to permit the lowering of the top blade guard to the work for thin plastics.
- Plan work for dead-end cuts to avoid having to back up work with the blade running.
- If possible (without too many people using the band saw), make a habit of releasing the blade tension after stopping the machine.
- Rip-cut along a marked line, by eye. The rip fence is often useless for straight cuts because of uneven sharpening or wear.
- If one side of the blade is very dull (causing the blade to lead away from the cut) it can be seen as shiny spots on the teeth of that side. Try honing down the other side slightly with a short pass against a sharpening stone to equalize the dullness.
- For tight corners and curves, use a narrow blade with the teeth set wide. With most machine instruction brochures tables are available that show blade widths vs. cut radius possible.

SANDING MACHINES

- Don't sand plaster on belts.
- Go slowly and lightly with the end grain of woods and with thermoplastics.
- With all disks, sanding removes less material near the center than near the edge. Slide the material along the disk for greater precision.
- Be sure to check the squaring (or angle) of the guide to the disk before working. The guide must also be tightly fastened.
- Glues are better (they are more secure) than other means of holding disks to the wheel.
- Clogged paper can be cleared with a wire brush or by sanding in reverse for a short time.
- Clean excess glue off joints before sanding their surfaces to avoid melting the glue and clogging the paper.
- Follow the arrows when replacing belts to prevent edges from catching and ripping the belt off at high speeds.
- A belt sander can often be made self-tracking by wrapping tape on one of the rollers (lower than the table) to make it bulged in its center. A slipping drive roller can be coated with neoprene glue for better friction.

THE DRILL PRESS

- Use the table height setting for a more positive depth stop if the machine's depth stop setting is weak.
- For precision holes, stop the machine when backing out the bit. Spade bits can bend while turning and create bigger holes than planned.
- Clamp a block at the hole's exterior for drilling angle holes to prevent the bit from bending and "walking" away from the center.
- Drill bits for abrasive, brittle materials such as acrylics should be sharpened to scrape rather than cut. A zero or even a negative angle is sometimes needed to prevent breaking out through the bottom of the hole.

- Use correct speeds and cutting angles (from common machine tables) for all sizes and materials.
- Strongly clamp the work to the table, in several places, for safety and precision.
- Center precision holes with a centering pin before inserting the drill bit, which may "walk." For long holes insert a centered pin into the table where the hole will be, to be sure it stays centered. Start holes with smaller bits to keep larger bits centered, but keep such holes small enough to keep the final bit point from wobbling.

THE MILL

- Feed the work against the turning cutting edge of the bit to prevent the operation from bouncing out of control.
- Firmly clamp the work in several places.
- When milling four sides of a block of wood, start work on cross-grain and finish with the grain on the last side. This prevents breaking off a good finished edge.
- If in any doubt about the machine's precision and quality, use the table height adjustment for a more positive depth stop.
- Feed the work slowly and take off light cuts for greater precision (since mill bits can bend into work as they turn).

THE LATHE

- For unfinished faceplate work, mark the center of the work on the plate, or keep the work attached if nobody else needs the machine.
- When sanding wood, use the reverse speed occasionally, to get rid of raised fibers.
- For taper or cam turnings, change centers once the work is round.
- Thermoplastic glue or glue with heavy paper is faster than screws for faceplate work.
- For inside cuts (bowl or cup shapes), reverse the turning direction or stand behind the lathe to turn for better control.
- Use shearing cuts for a better finish and less sanding.
- Use templates (with the machine stopped) to check precision or repeated work.
- For cylinders, mark and turn to diameter at regular intervals, then connect with a straight turning. For curved work, intervals should be closer on tighter curves, for better control.
- Cutting square stock octagonally before making it round is not worth the trouble. Use a heavy, sharp gouge to rough shape a round shape, keeping the gouge on its edge along the tool rest for fast shear cutting.
- Keep the tool rest adjusted close to the work (about 3 mm.) for safety and precision. Adjust the height to your own comfort (approximately along the centerline of the work) and for where you want the chips to fly.
- If the dead center is not a ball bearing, keep it lubricated with wax or oil to prevent burning the center hole bigger.
- Leave plenty of waste at every planned cut-off section when marking work, to make turning easier and more precise.

THE GRINDER

- Don't grind aluminum, wood or thermoplastics against the wheel, as they will quickly clog it.

- Don't sharpen tools in a fireworks of sparks—the heat will destroy their temper. Use brief, light passes, and cool the tool occasionally in water.
- Don't use the felt wheel for removing the burr after sharpening. It may feel right, but it is not a good cutting edge. Exercise patience and use an oilstone by hand.
- When using a felt or cloth wheel for polishing plastic edges, don't start at the end of an edge. The corner may catch in the wheel, pulling the work out of your hand.

THE JOINER-PLANER

- When planing all four sides of a wood board, start cross-grain and finish with the grain to prevent breaking through the finished edge on the last pass.
- With bent work, start by planing the concave side.
- Plane with, rather than against, the grain whenever possible. This avoids splintering and rough surfaces.
- Don't attempt to plane small pieces of work.
- Keep a constant follow-through of downward pressure on the work to maintain a flat cut.
- Don't expect a constant thickness along a piece of work with this machine.
- The outfeed table must be perfectly adjusted (to the maximum height of the turning blade) for a flat surface. Infeed and outfeed tables must also be perfectly parallel. If they are not, use shims to correct the alignment.

THE THICKNESS PLANER

- Pass the work on the joiner-planer first to get one (or two) truly flat sides. A rocker shape that has only been thickness-planed will remain a (thinner) rocker shape.
- Lubricate the feed table often (with kerosene on a felt pad glued to a flat stick, with the machine stopped), especially for narrow work.
- Slightly hold up long, heavy work at the beginning and at the end of the cut to avoid the dent often produced by planers at work extremities.
- Don't try planing short work.
- With wood, at least the final pass should be with the grain.

THE SABRE SAW

- The sabre saw may be used upside down to better see the traced line through all the dust it produces.
- The machine's shoe may be removed and replaced by a table to use the machine as an upside-down jigsaw.
- Use special blades at reduced speeds for cutting metals and plastics.
- The sabre saw can pierce material, but practice this cut on scraps.

THE ELECTRIC DRILL

- Ask for help lining the drill up to the correct angle, which should be perpendicular to your view, as well as across your view.
- Work at a comfortable height and with good support.
- Stop the drill before entering a started hole (after removing it to clear chips) to prevent chattering and ruining the hole.

HAND TOOL SAFETY

Many of the safety rules for using hand tools seem to make good common sense, but accidents do happen, even with the simplest tools. Always clamp material and aim screwdrivers, chisels, gouges or cutters away from your body or away from anyone nearby. Bare feet and sandals should never be worn while performing any of the aforementioned procedures, and hand tools are no exception. Only tools that are clean, sharp and in good condition should be used, and match the right tool with the job; a rubber disk around a chisel may prevent hammering knuckles, for instance, and wrenches of the correct size for the given nuts or bolts prevent slipping and knuckle injuries. Keep a special container for the sole purpose of disposing of cutter blades. Should these get left lying around, or even thrown in a trash bin (where modelmakers sometimes scavenge for material), someone could get hurt. When starting a cut with a hand saw, pull first to enter the blade into the wood safely (the exception here is the classical Japanese saw which cuts on pulling). At the end of the cut, use less pressure to avoid cutting yourself with the free blade. Wear safety glasses when there is a risk of flying debris.

If you have an accident—whether you're an expert or a beginner—stop working and get help. A small cut will not always require stitches, but medical decisions may need to be made to avoid infection and to facilitate quick healing. Always consult a doctor if you think you have an eye injury or sensitivity.

MAKING SPECIAL TOOLS

Most of the time modelmakers and craftsmen will buy their tools, sometimes accepting only the very best available. Once in a while, though, it may be useful or even necessary to make some special tools, in hardwood, plastic, iron (mild steel), hard steel or other metals. In some countries good tools are either very expensive or simply unavailable. This is often the case for drawing sweeps, transparent drawing bridges, good styling clay tools and even simple, adjustable plastic mannequins in a desired scale or "percentile" size. Other simple tools that can be hard to find are transparent center finders, hooked scrapers for acrylic sheet cutting, heavy, wide squares for clay work, clay and foam scrapers of a certain curvature, custom templates and oversized calipers or compasses. It can be easier and faster to make a tool than to go looking for it, especially if the project is a unique one, after which the tool may be discarded. What follows is a description of techniques that can be used to make many types of simple tools, including some of the ones mentioned above.

WOOD

The wood used for tool handles, clay scrapers, etc., is usually a hardwood, selected for durability. If it is being used for scrapers, it must also be a dense, homogeneous wood. Apple, pear, boxwood, sycamore and sugar maple, and to a lesser extent, beech, cherry or walnut are examples.

Hardwoods such as oak, ash or elm have too much pronounced grain to be used for good, smooth styling scrapers.

But even hardwoods have their weaknesses—in this case, a tendency to split open. For this reason, handles for clay tools, chisels, files and gouges must be reinforced with tight-fitting steel collars, or be tightly wrapped with wire and glue on the end.

Rough work to begin wood tools can be machine done, using a lathe (for handles) and power cutting and sanding. The final sanding should be done by hand, for better control of shapes and scraping edges. For specific curve shaping, parts of the model's drawing can be traced and reproduced on a scraper. Figure 14-7 shows a few examples of wood (or hard plastic) scrapers. If well-made, they may be worth keeping around for future projects, even if they have been custom-shaped.

Fig. 14-7. Wood and plastic scrapers.

PLASTIC

Hard, transparent plastics can often make useful tools. The plastic tools mentioned earlier in this chapter are simple enough to make

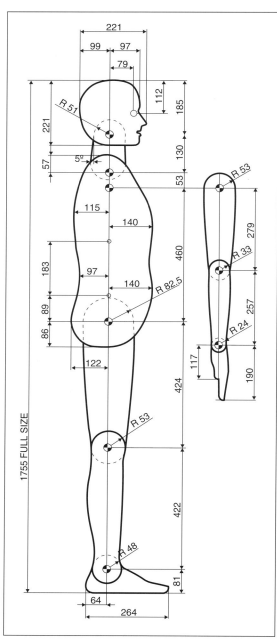

Fig. 14-8. A template for an adjustable mannequin.

without any further descriptions here. In making a mannequin, the outline shown in Figure 14-8 may be traced and reduced or enlarged to the scale or percentile needed. A percentile number for a specific-sized mannequin corresponds to the percent of similar-sized public who usually will buy the product being designed. For example, a 60 percentile auto mannequin represents the average size of 60 percent of the people who usually buy that car.

A FEW OLD TOOLS

Long before design was even a word, hand tools were made for specific jobs. Often the craftsmen made them for their own projects and had big collections of tools for many purposes. Live oak (very hard), brass, hard steel and the like were often combined in a single tool. A few examples of these are the molding planes of woodworkers, adjustable planes for convex or concave curves in boatbuilding, oversized compasses for barrelmaking and finely-engraved handles with diamond chips for glass cutting. Woodworkers and blacksmiths, especially, had the greatest variety of handmade tools.

IRON AND STEEL

Metals and their alloys are widely used in modelmaking today. Some of the operations include casting, welding, machining, bending, swaging, heat-treating, forging and chemical etching. The information presented here is limited to some theory and techniques that may be most useful in a model shop. For example, clay shaping tools may be made fairly quickly in iron or hardened steel.

Pure, or soft, iron is almost always mild steel—iron with a small percentage of carbon—which cannot be hardened by heat treatment. This mild steel, from heavy nails, for example, may be hard enough for some temporary tools. It may also be faster to do the frequent sharpening needed on this steel than to make a hardened steel tool.

But sometimes hard steel—with a higher carbon content—is preferred. Properly made and heat-treated, such a tool will last a long time and its edge will stay sharp, even with frequent use. The heat treatment required consists of hardening, annealing and tempering.

HARDENING

High-carbon steels (and alloys) may be hardened in two ways:
• By quenching, or plunging the red-hot steel into a cold liquid (water, brine, oil, etc.).

- By working, which can be bending, hammering, etc., of soft (annealed) steel.

Hardened steel is brittle and can easily break. Viewed through a microscope it looks like a rough, crystalline structure. This results from rapidly cooling red- or white-hot steel, or from working softened steel cold. A quick experiment: bend a steel paper clip out of shape. Try to bend it back to its original shape and it will bend elsewhere more easily, because the first bending has "worked" the steel, hardening it more than other parts. If you do succeed in bending the steel in the same place again, it will get even harder there and soon break apart. This action of work-hardening steel is an industrial method of treating some tool parts, but it can be an inconvenience while forming a tool. This hardness must be eliminated by annealing while making the tool. Such hard steel not only breaks easily, but may often be even harder than the (tempered) steel of files, hacksaws and cutting pliers, making their use impossible.

ANNEALING

High-carbon steel may be softened to make it workable—for a while— by annealing. This softening process simply involves heating the steel red-hot, then allowing it to cool very slowly, preferably in a bed of insulating material such as ashes. Such soft steel, if examined with a microscope, would appear smooth and homogeneous in structure, with none of the crystals of the same steel in its hardened state. Annealing is necessary for the shaping—by hammering, bending, drilling, filing, etc.,—of a tool. And since this shaping will work-harden the tool's steel, it must be stopped before it is too late, and the annealing repeated to make the steel workable again. It may be necessary to repeat this process over and over if you're doing a lot of work on a tool to shape it. But that's preferable to starting all over again if it breaks.

TEMPERING

Tempering was the art of the blacksmith's work. Today it remains a complex process, with many variations and types of metals produced. Steel, if only quenched or work-hardened, is too hard and brittle for use as a tool. It is also too soft if only annealed. Tools require varying degrees of hardness and flexibility between the two extremes, and tempering produces the correct hardness.

Files, for example, must be very hard in order to file iron and steels. Knives, scrapers and screwdrivers must have hard, but not brittle, edges, yet must be flexible or springy in their bodies or handles to prevent breaking. A good clay scraper must be tempered hard enough to keep its cutting edge without frequent sharpening, but must resist the bending it undergoes as it is pulled repeatedly through clay.

The many tempering processes, some used for centuries, treat different tools differently to make them both useful and long-lasting. What follows is a description of one such process.

Once the tool has been shaped completely (except for final sharpening and polishing), it is ready for tempering. First, to harden it, it is heated red-hot and quenched. Then it is filed or sanded to remove the oxide and make it shine. Next it is heated again. The shiny steel will begin to change color, starting with the hottest part. As it is heated it will turn from various shades of yellow, to purple, and finally to blue before becoming red-hot. These colors correspond to the changing hardness of the steel as it is reheated.

Various tools require different hardnesses. For example, the yellow range yields the hardest steel (next to being brittle), used for metal cutters and razors. Purple produces a spring temper. Dark blue is good for some saws. Heated to light-grayish-blue, the steel is too soft for most tools. It is then no longer tempered, but almost annealed. For a clay tool, somewhere around dark yellow or purple will provide a good temper.

Tempering, then, may be seen as a controlled reheating of a hardened steel to remove some of its brittleness. Some of the more complex tempering methods include hardening cutting edges of knives while keeping the rest of the blade soft, or hardening the exteriors of gears with soft interiors. The simple approach described above should be enough to make most of the simple tools used for special modelmaking work.

MATERIALS AND EQUIPMENT

Assuming the project consists of a few custom-made styling clay tools, these can be made from readily available steels with a minimum of equipment. Piano wire and spring steel bands can be bought in various thicknesses. Worn hacksaw blades or files can also be recycled. Lengths of hardened steel, as-bought, can be "cut" by simply bending off a piece in a vise with a hammer (using work-hardening to your advantage, instead of ruining cutting tools or trying to anneal big amounts). Wooden dowels or a lathe will be needed for handles, and a length of steel tube to cut into rings to reinforce them. Equipment should also include safety glasses, leather gloves, a strong vise—preferably with an anvil—hammers—including a ball-peen—pliers, files, a butane or propane torch and silver solder for brazing if strong joined parts are needed. When heating to braze, heat the metal part until it is hot enough to melt the silver solder. Just heating the solder to melt it onto cooler metal will result in a weak "glue job." Don't clamp the part in a vise while heating it—use old pliers. The metal vise will absorb all the heat applied to the part, which will never get hot enough for brazing, annealing or quenching. Again, be sure to stop work, as often as necessary, to anneal a part that has hardened by working it.

A final polishing and a strong, reinforced handle will complete the tool, which you may want to show off (then hide, with your initials engraved, in your tool box).

CHEMICAL ETCHING

The process of chemical etching is used by artists, craftsmen and in industry to both engrave metal surfaces and to remove large amounts of metal in a controlled way. Tool blades may be etched in relief with names or logos, and the aircraft industry "shapes" solid parts into hollows for lighter weight.

Three of the most common chemicals used to etch metals are:
- Nitric acid, for iron and steel.
- Ferric chloride, for copper.
- Sodium hydroxide, for aluminum.

CAUTION—THESE CHEMICALS ARE VERY DANGEROUS. They can cause severe burns or blindness. Any basic chemistry book can refresh your memory on how to use these substances safely. If in any doubt at all, just don't use them.

The etching action on metals is a chemical reaction. The liquid eats away at the exposed metal, which becomes a salt, soluble in the liquid. The longer the exposure, the deeper the etch. Masking, to protect parts of the metal from the chemical, may be done in various ways, using materials that resist attack by the chemical used. For much surface etching, a special liquid tar may be bought. It is brushed on the metal surface and allowed to dry, then scraped away where the metal must be exposed to produce the engraving. This will produce a direct relief. The masking liquid may also be penned or brushed on only the parts to be protected, to produce a background relief etched away. Liquid, "peelable" rubbers may also be bought that are easily removed from the surface once dry. A sharp cutter is used to cut the dried rubber skin, which is peeled off around the cuts to expose the metal for etching.

The masked metal is placed in the chemical bath and allowed to remain long enough for the depth of etch desired, then removed and rinsed, and the masking material removed as well. For very deep etches, the chemical will create "undercuts" beneath the masking by attacking the metal in all directions once it has gone down beneath the masked layer. Again, etching chemicals can cause severe damage and must be handled, stored and disposed of with special care.

Figures 14-9 and 14-10 show a few styling clay tools made using the techniques described in this chapter.

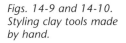

Figs. 14-9 and 14-10. Styling clay tools made by hand.

Photo, design and model: Norman Trudeau.

SANDING AND FINISHING

USING LIGHT AND ABRASIVE PAPERS

Since they occur near the end of the work on a model, sanding and finishing mean working with smooth surfaces and scratch marks that are not always visible in most light. Interior lighting is usually too poor for good sanding, and sunlight is often too much of a good thing. Outdoor light, especially the good old north light of painters, is what works best.

When reading or writing, rear lighting—a light that comes down over your shoulders—is often recommended. This prevents glare on the paper, which can eventually tire the eyes. With sanding, the best way to see surface scratches and other tiny imperfections that need further sanding or filling is to look right along a surface into the glare. This way, even small scratches will show shadows. Rotating the model, or moving around as you look into this light will help catch all the variations of light reflected toward you from the surface. For bigger, softer defects, feeling with the fingers can also find unfinished work. One art school in London even asks that students sculpt and finish an egg while blindfolded.

Dusty as it will be, the surface must be wiped or blown clean often, to reveal the work to be continued. In addition to providing the best light for sanding, the outdoors, or at least the proximity of a good draft from open windows and doors, is also better for the health. Sanding produces more dust than almost any other type of operation. A good dust mask should always be worn while sanding, and holding the model downwind will also help. Hand sanding also produces a lot of dust in the air you breathe. Some of this dust is particularly harmful, as in the case of epoxies and polyester resins and fillers. But all dust may be harmful if breathed often and for a long time. Stone dust and coal dust often cause forms of silicosis, a permanent industrial ailment that took 30 years to be recognized by United States medical, industry and government leaders, after having been proven in Europe. Bakers and woodworkers sometimes have to quit their trades for good because they have developed a sensitivity to these "harmless" dusts. Urethane foam is also considered by its manufacturers to be "inert". This supposedly means harmless, but many people get a skin rash when exposed to urethane dust. Some "safe" dusts, like powdered lime, are so fine that they produce immediate coughing when in the air, and can be the cause of a quick trip to the hospital for some people. Occasional sanding is usually considered harmless. But anyone doing it (or anyone exposed to it) for extended periods of time, such as students, teachers, cleanup crews, shop personnel and designers, should seriously consider the risks.

Toward the end of the sanding, it may help to spray a very thin coat of fast-drying paint on the surface in a contrasting color. Then lightly sand it all off! Doing this will not only fill in some tiny scratches, but will also reveal bigger defects. A sanding support behind the abrasive paper (don't just use the fingertips) will allow the paper to bridge over pits in the surface and will sand bumps down. The pits will be visible as color contrasts. Then a decision has to be made: to achieve a smooth, continuous surface, either sand the surrounding surface down (which may result in a design change if there is a lot to sand off) or fill the pits and sand the filler when it dries.

There are many kinds of sandpaper available. The grain size is an indication of the sieve through which the grains were sifted to be glued to the paper (or cloth). The higher the number, the more divisions in the sieve, so the finer the paper. Today, a grain size as fine as 1500 is available for pre-polishing work on plastics, metals and gloss-painted surfaces.

When a model must be shaped and fine-sanded, it is a mistake to start the sanding with a fine-grained paper. Time and paper can be saved by rough-sanding at first with coarse paper (this can be as low as no. 40), and progressing to the finish in stages (for example, 40, then 60, 80, 100, 120, and so forth). The trick is to eliminate the scratch marks of each previous sanding grain with the next paper used. Once the surface is covered with the new, finer scratches, go on to the next finer grain.

It is also a mistake to sand a surface too fine if it is to be painted. Paint, and its primer, must cover and fill the final sanding scratches. But a surface may also be sanded too smooth to hold paint well. Fine scratches (600 to 800) are often needed to mechanically hold the dried paint to the surface.

The paper may be "wet"—made with glue that will not soften in water—or "dry." Many kinds of abrasives are used to make these papers. Classical sand and glass are harder than most materials. The hardness is necessary, otherwise the grains would wear down faster than the surface being sanded. Even harder materials, such as silicon carbide and aluminum oxide, last longer. The shape and spacing of the particles may also vary. Some are splinters or hook shapes, spaced farther apart on the paper to prevent clogging up. Finally, there is the quality of glue and paper used. Some abrasive papers are so good that the others are just a waste of time and money.

Usually, cloth backing is needed for belt machine sanding or hand operations which would likely tear paper. For power disk sanding either kind can be used, but an open-grained, resistant paper will last a lot longer.

HAND AND MACHINE SANDING

Most design model sanding is done by hand, although there are always some who take chances with a belt sander, preparing four or five models just in case.

Hand sanding can involve a great deal of time-consuming work. This is always justified for small, curved work, where constant control of the surface is needed. With big, flat edge or surface sanding, time can be saved with machines, including a planer.

But there are reasons for avoiding sanding machines. For example, the usual belt or disk sanders take away so much material so fast that the sanding may easily get out of control. Both sanders may also dig into the surface with their edges, and disks may be destructive, both on flats and curves. For finish work, vibrating sanders are safer, but still require practice for good results. Outside curves are difficult enough, and inside curves are almost impossible, unless done by hand. Drum sanders for some thin, inside contours may be used, but these are also hard to control for finish sanding.

Another important reason for hand sanding is the friction heat created by machine sanding. All thermoplastics can melt with heat, and will under heavy pressure and high speed with a sanding machine. Even final polishing of plastics such as styrene and acrylic must be done slowly and with care if a machine is used.

Finally, there are the fingers. Sanding machines are not especially dangerous, but they can catch loose clothing, jewelry or long hair in their moving parts, and can very easily sand the fingers. Such a wound, although usually not terribly serious, is very painful and can take a long time to heal.

SUPPORTS AND SANDING MOTION

For most work, sanding supports should be used behind the paper to act as a bridge. The fingers will not bridge over wide areas. In sanding bumps or pits on a surface, the fingers tend to follow the imperfections,

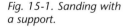
Fig. 15-1. Sanding with a support.

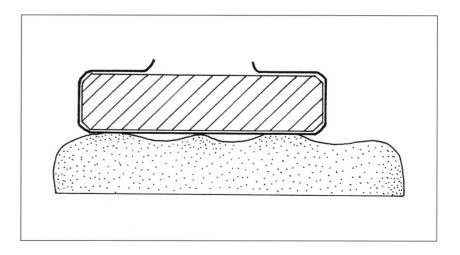

Fig. 15-2. Producing a rocker.

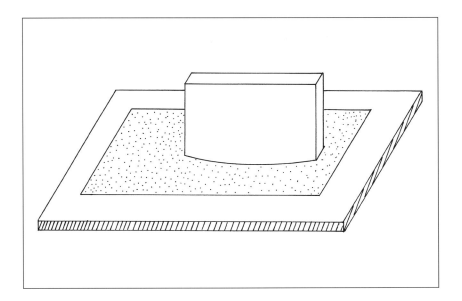

riding up and down like waves in the ocean. This results in a removal of the surface, but the bumps and pits remain. The bridge allows the paper to span over such defects, and either sand down bumps to the surface, or sand the surface down to the pits.

Supports should not always be hard. Flat wood blocks, for example, are needed to sand edges and flat surfaces, without the rounding effect of softer materials. But cork, rubber, felt or soft foam should be used where a flat, hard edge would show sanding marks. Hands and fingers may also be used for curved work, where bridges would easily create flat spots on convexes, or not fit into concaves smoothly.

Motion in hand sanding should vary depending on the work. When sanding edges on acrylic sheet, for example, care must be taken not to produce a rounded edge, which will not fit later for precise construction or solvent bonding. Figure 15-2 shows the common result of sanding along an edge without care.

Abrasive paper can be taped to a smooth, flat table, and the plastic sheet pushed slowly back and forth to sand along its edge. Care must be taken to keep the pressure even, and the motion slow, to avoid the rocker shape which may easily result from putting more pressure on the ends, as seen in Figure 15-2. When sanding such an edge, even finish sanding of the flat edge should not be done with just the fingers. This would press the paper more along the outside edges, and the edge would be rounded, as seen from the end. Such an edge is difficult to solvent bond.

SANDING WOOD
Wood requires special treatment when sanding because its fibers lead in one direction. It is easy to scratch and rip these fibers by sanding across them, and this would create visible marks. Sanding along the fibers takes longer, but leaves fewer visible scratches which tend to "blend" with the grain of the wood. The rule, then, is:
- Rough sand across the wood fibers, for faster work.
- Fine sand along the fibers.

For most woods, a grain of 120 to 180 paper is usually fine enough in the final stage of the process, since such scratch marks will be invisible and are no rougher than the wood grain itself. Sanding end grain, or the cut off fibers, can be difficult. These fiber ends resist abrasion (that's why some good hardwood floors and butcher blocks are made with exposed end grain). If there is a temptation to save time by machine sanding end grain, the friction heat built up by the wood's resistance may be enough to quickly burn and darken the surface. Sometimes a drop of water (lick the finger and rub it on the wood) will cool it off for faster sanding. This is not a good idea where precise fits are needed, because the heated water will swell the wood. The best thing to do is to simply take the time needed to do the job properly: short, light sanding applications, with little pressure, and frequent pauses to let the wood cool.

METAL, PLASTIC AND PAINTED SURFACES

With homogeneous materials like metals, plastics and paint, there are no fibers that require a special sanding direction. But there is another reason for changing the direction of sanding motion, including with wood. If the surface has waves or ribs, and the sanding support is soft enough or small enough to follow into the hollows under pressure, sanding in only one direction will continue these contours as material is removed. This can most often happen when shaping or sanding large surfaces with slow curves. What is needed is a sanding support large enough and firm enough to bridge across these big waves, and to change sanding (or scraping) direction often.

For thermoplastic surfaces, again, machine sanding must be done with care. Thermosetting plastics will not melt with sanding heat, but are usually very resistant to abrasives and take a long time to sand. In general, these homogeneous materials can, and should, be sanded much finer than wood, since their nonporous surfaces will become smoother with each finer grade of paper used. The important limit, as mentioned above, is to avoid sanding the material too smooth for priming or painting, preventing the dried coats from adhering. For heavy fillers, such as polyester spray filler or paste filler, the final sanding may even be as rough as no. 80. This represents a big time savings, and a good rough surface for the filler to cling to. For most spray lacquers, no. 600 to 800 is usually fine enough. The finer papers, wet or dry, are used to sand dry gloss paint or for surfaces not destined for painting, as preparation for polishing.

POLISHING

Polishing metals and plastics may be considered extremely fine sanding. Polishes in liquid or paste form exist in several grades (from the grains they contain) for preliminary and final luster polishing. Machine polishing is done with cloth or felt wheels for faster work, but again, this may be too fast for thermoplastics. For transparent plastics, it is difficult to take out scratches and return to the original shine. Trying to polish out scratches from plastic eyeglass lenses is one example of this. Even on much bigger and more expensive transparent surfaces, such as aircraft canopies, technicians will often replace the part rather than try to sand down and polish deep scratches. So much material must be removed that it would create dents and distort the pilot's vision.

WET SANDING

Wet sanding could often be seen as the scourge of modelmakers, and some wet sanders should probably be prevented from being anywhere near others, who are trying to avoid problems. Dry sanding is bad enough, for the health and for drying paint, because of the dust it produces. Wet sanding, though it does keep the dust down from the air, offers special problems.

Sure enough, wet sanding saves paper, because it lasts longer when it is lubricated and rinsed off during the sanding, especially with slightly soapy water. This wetting also leaves fewer scratches on a surface in the final sanding stages, where one dry, trapped grain from a previous paper may be rubbed around to damage a smooth surface. The shine produced by the water film also helps show the highlights and surface irregularities. On wood, slightly dampening the surface before sanding will raise the surface fibers as they dry. These, once sanded off, result in a more permanent final sanding.

Some of the problems that come from wet sanding are serious. Just a little water absorbed into an apparently waterproof model surface may cause big problems if the model is to be painted. The surface may seem dry, but most often it has not been given enough time to dry completely. Once the paint is on and dry, this water will want to come out, and that usually means blistering, peeling paint. Silicones, too, are dangerous to paint jobs if they are used anywhere in the vicinity (many auto body shops have banned silicones for that reason). Just a few molecules of silicones in the air can cause "fisheye" on painted surfaces, and this can be hard to correct. Such silicones may be part of the polishes that some wet sanders actually use on their models before painting, or rather, trying to paint. Or they may remain as residues in containers used to hold water by other wet sanders. This happened to one former student, and his entire fifth-scale car model was one big fish eye. Wet sanding is also often done to such a fine grade of paper by some believers that even if the surface is totally dry and free of silicones for the paint job, it is just too slick and shiny to hold the paint.

For those who want to see their models shine, wet sanding should be saved for after the paint job, or on unpainted surfaces. This should also be done far from any influence on susceptible models or modelmakers. The paint must, of course, be completely dry before final sanding, wet or dry, and polishing.

Fig. 15-3. Soapy water and wet and dry paper. Photo: Monique Carroll.

Fig. 15-4. A blistered paint primer coat. The model part was still too damp from wet sanding to be painted. Photo: Monique Carroll.

Photo, design and model: Ulrich Schraudolf.

CHAPTER 16

SURFACE TREATMENT AND PAINTING

SURFACE TREATMENT

SANDING

The surface to be painted must be brought back from a too-smooth finish if it has been sanded too fine. Paint and primer hold best on a surface that would look rough through a magnifying glass. The scratch marks from the final sanding should be just fine enough to be covered over by the coats of primer and paint to be applied. This can vary from thick, viscous polyesters and primers, where 80 to 100 paper is fine enough, to thin lacquers, where 600 to 800 paper may be needed—but no finer than that. Save the super 1200 and 1500 grain, and especially the wet sanding, for polishing preparation of dry, painted surfaces.

CLEANING

All dust, wax and grease (from perspiring hands, for example) must be removed from the model's surface. Wiping it lightly with alcohol and blowing it with compressed air is usually sufficient. If the surface is wet or even damp (from rain, sneezing, condensation on a cold model in a humid room or especially wet sanding) the model must be dried—and this may take days. Do not force drying with a hot air gun, as this may cause the paint to blister. Give the paint a chance to stick, at least for the presentation of the model, and use a tack rag for a final wipe of the surface if necessary.

PRIMING

Primer, as an undercoat before top (color) coats, serves two important purposes: it fills tiny pores and scratches on the surface better than paint because it is thicker (and less expensive), and its composition allows it to stick, not only to the surface, but to the paint which will follow. Some primers or pre-primer washes have a special chemical action, such as the phosphoric acid-based liquid used on aluminum aircraft to pickle (etch) the surface for better paint anchorage. There are also primers and paints with solvents that slightly dissolve a specific plastic surface, then evaporate fast enough to do no damage to the surface. This results in a virtual solvent bond of the paint to the surface.

MASKING

Unless absolutely necessary, it is always better to avoid masking. If a model is multicolored, and masked to paint the various colors, the masking lines almost always show. To avoid this, and to have it look more like the real product designed, a clean boundary must exist between the different colors. This is almost impossible to acheive through masking. For example, if two colors on a model represent separate parts assembled at a cut line, then the model can be made in separate parts and joined after painting.

If masking is necessary, the following rules will help avoid at least some of the big problems:

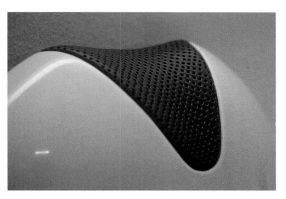

Fig. 16-1. A paint job in separate parts. Photo, design and model: Vasilis Milonadis.

- Use low-tack masking tape. If it can't be found, get rid of some of the tape's stickiness by rubbing it along dusty clothes or greasy fingers. Tape that sticks too well may pull off a first dry coat of paint.
- Allow the first color coat enough time to dry before applying the tape. With most acrylics, for example, this means at least a day in good drying conditions (watch temperature and humidity). If the first coat is not dry enough, it may pull off the surface when the tape is removed.
- Rub down the tape along the edge to be painted to prevent paint from creeping underneath, resulting in a jagged color line.
- Once the second color is sprayed, let it dry long enough to prevent it from stretching off the surface when the tape is pulled. Leaving the tape on too long, however, will allow its adhesive to set too hard, and may pull off the first coat when removed.

PAINT PREPARATION

Consider all factors that are relevant to the construction of your model when selecting the paint—drying time, gloss required, painting conditions, etc.. Buy the primer with the paint from a knowledgeable salesman—a vanishing breed because of the many types of paint available today. Read all the instructions, including those on safety and health risks, and follow them. If a certain viscosity paint is needed, use it, even if it means running the mixed paint through a viscosity measuring cup. With experience, this will become easier, and you will be able to see whether a paint runs down a mixing stick like water or like milk. Mix with a clean stick, and filter the paint through clean filters into clean cans for storage. Keep the cans covered and labeled. For two-component paints with a short pot life, also label the cans with the time and date of mixing the catalyst. It is surprising how much paint has to be discarded because of rushing and carelessness. But when discarding is really necessary, do it safely. Enough harm has already been done to present and future water supplies by pouring poisons down the drain.

AREA PREPARATION

There is no use trying to spray in the model shop, or even in a spray booth that is pulling in air full of dust, fiberglass strands, silicones and water vapor. It would be better to take a chance with the insects outdoors. An air-solvent extractor is needed in the spray booth for both health and paint quality reasons, but the air being pulled in to replace the extracted air should be clean, warm and dry. The best is a system that draws in heated and filtered air faster than it is extracted through the spray booth ventilators. This results in a positive air pressure in the spray area which will push out unwanted air (instead of letting it in) every time a visitor or a sprayer opens the door. Clean the area if it is dusty, even as far as vacuuming the walls and washing the floor. Blow yourself and your model clean with compressed air, but make sure to do it somewhere else.

TYPES OF PAINT

Paint is essentially a mixture of pigment (an insoluble colored powder), a binder liquid that dries to form a skin containing the pigment and a solvent that dilutes the liquid to a workable consistency. As the solvent evaporates, the paint drying takes place. The binder also hardens to complete the drying process. With the two-component paints available today the binder dries as a result of a catalyzed chemical reaction.

NATURAL OIL PAINT

Oil paint has been around for centuries. The binder is linseed oil, from the linen plant, and the solvent, which used to be turpentine from pine trees, is now more commonly derived from petroleum products. Today it is hard to find this paint except in tiny artist's tubes, yet it remains a proven favorite of experienced craftsmen and house painters all over the world. Oil paint is slow drying and usually applied by brush or roller. On a well-prepared surface it holds so well that it can last for hundreds of years. It is not often sprayed.

WATER-BASED EMULSIONS

Emulsions are whipped-up mixtures of water, pigments and a synthetic binder, such as liquid acrylic or vinyl plastic. The water makes it a healthy substitute for solvent-based paints. It is medium-fast drying, and used for building interior and exterior work and on some model surfaces, like polystyrene foam, which would dissolve in most paints' solvents. It can be sprayed, but is usually rolled or brushed.

SHELLAC

Shellac is a clear (no pigment) solution of alcohol (wood or ferment) and a natural plant resin or gum. It is medium-fast drying and, like emulsions, will not dissolve thermoplastic materials such as styrene and styling clay. For centuries, it was rubbed on in many coats to produce a high gloss finish on fine furniture, a process that demanded a great deal of practice. Today, it is simply brushed or rolled on, but is used in modelmaking only for special purposes, such as protecting styrene foam or clays from other liquids.

LACQUERS

Classical lacquer is a clear cellulose plastic binder dissolved in a fast-drying solvent, with or without colored pigments added. It is most often sprayed and produces a hard, glossy finish. It is often replaced today by acrylic lacquer, a blend of liquid acrylic resin with a variety of solvents chosen for their drying speeds. Acrylic is softer and more elastic than cellulose and will not crack as brittle cellulose does with time.

ENAMELS

Enamels give extremely hard, long-lasting surfaces, but are very slow drying unless "baked" in high temperature drying rooms. They are seldom used in modelmaking.

TWO-COMPONENT SYSTEMS

The two-component paint systems are usually thermosetting resins, such as polyurethane, polyester and epoxy. Such resins harden by a chemical reaction called polymerization, which takes place when a catalyst is added to the resin (although it can happen by itself over time). These paints give good results after some practice, but they can also be risky. Forgetting a mix in the spray gun for longer than the pot life of the catalyzed resin causes it to harden. Once hardened, the resin cannot be dissolved in most solvents and the gun must be thrown away. The resins also tend to dry so hard that polishing to even out surface defects becomes very difficult.

Urethanes are often used on floors and tables to resist wear. In modelmaking they can be useful in painting PVC surfaces, which typically do not hold paint well. Polyesters and epoxies are also very hard when dry.

THE SPRAY CAN

This is, of course, not a kind of paint, but a type of packaging. It is a do-it-yourself approach to spray painting for everyone, including the modelmaker. Nevertheless, the quality of work with spray cans is very hard to control, and they are mostly used for fast jobs or very small parts when it is not worth buying and mixing enough paint for a better spray job with a gun. It is important to be careful when buying spray can paints. Many of the above-described paint types are sold in spray cans, and in some places, where ingredient labeling is not required of manufacturers, it is common to see labels such as "Synthetic" or "Contains Chemicals," which constitute no information at all. If you are in a hurry to finish, for example, you really don't need a slow drying enamel. And it is very important to buy primers that are compatible with the paints that will go over them, and stick to them rather than dissolving them into a slush. Some of these unknown chemicals may also dissolve the model material itself.

PAINTING TECHNIQUES

All of the known techniques—spraying, brushing, rolling, dipping, etc.—are used to finish design models. The following information is concentrated on spray painting. However, many of the points discussed will also be useful with other methods of painting.

SPRAY PAINTING

There could be a lot to write and study about spray painting, especially with the evolution of paint chemistry over the past few years. There is also a lot of practice involved—years, maybe—to get really good at it. But even with all the variables, a little understanding can be a big help. Spraying can be a lot of fun, with many satisfying results. It can also be a frustrating, unhealthy experience if care is not taken.

THE PROCESS AND ITS VARIABLES

With most spray guns, the paint mixture is blown from the gun by compressed air, resulting in a fine mist that covers the surface to be painted. There are various types and qualities of spray guns. The only aspects of spray guns that will be discussed here relate to the typical adjustments necessary for spraying. These, on a reasonably good gun, are:
- Paint feed.
- Paint-air feed.
- Nozzle orifice size and position.

Other important variables, independent of the spray gun, are:
- Air pressure.
- Temperature of the air in the spray booth.
- Humidity of the air in the spray booth.
- Temperature and humidity in the drying area.
- Paint type.
- Distance from the spray gun to the surface being painted.
- Speed of the air extracted through the spray booth.
- Cleanliness of the model, the area and the sprayer.
- Dryness of the model.
- Humidity in the compressed air to the spray gun.
- Speed of the gun's movement while spraying.

These variables not only affect the spray job, but also each other. Often, correcting one of these will allow or force you to change another for good results.

CONTROLLING THE VARIABLES FOR GLOSS SPRAYING

The paint mist droplets, blown by compressed air, should land on the surface as a liquid. The droplets will then merge to form a liquid film on the surface just before the solvent begins to flash off (evaporate). The following techniques will help this happen as it should.

The spray gun must be held at the correct distance from the surface. Usually, about 8" (20 cm.) is a good average. Too far, and the solvent will start evaporating before it hits the surface. The result will be sticky, partly dried droplets that will not merge to form a film, but remain as a bumpy "orange peel" surface. A greater distance means even drier droplets deposited on the surface as a chalky paint dust, called "dry spray." The correct distance will vary, depending on the mix viscosity, air temperature and air flow in the spray booth, air pressure from the gun and the type of paint used (for example, the paint's solvent may be fast or slow to evaporate). It may also be useful to insert a "holding stick" into the base (or another area that will not be visible at the presentation) of the model. The model can then be moved and turned while spraying, as well as the gun. Such a stick will also allow a small model to be sprayed at one time, and can serve as a drying support.

Aim the spray gun at the surface so that the paint mist hits the surface head on, with the spray perpendicular to the surface. Otherwise, an uneven concentration of mist is deposited and part of it will be drier, due to the greater distance it travels in the air.

Practice moving and aiming the spray gun. On large, flat surfaces, it means moving back and forth evenly along the surface, applying overlapped coats, and keeping the gun not only perpendicular but moving it parallel to the surface at the correct distance. Spraying curved surfaces is more difficult, but try for the same approach, always aiming the gun and following the curved surface at the same distance. This may take some fancy body work.

If all goes well, the dried coat of paint on the surface will require little or no polishing to achieve a smooth, glossy finish. The last coat sprayed on is especially important. Some lacquers will need five to ten coats to completely color the surface. Other paints will cover with only two or three coats. In most cases, though, don't even try to color the surface with the first few coats.

Depending on the position of the adjustable nozzle, paint mist is deposited on the surface (by a stationary spray gun) in one of the patterns shown in Figure 16-2.

To even out heavy and faint distribution of paint in these patterns, overlap the coats while moving the gun. On large surfaces, a combined back and forth, then up and down overlapping helps give a more uniform coat. On any surface, because of this uneven mist deposited, and because the paint must be very liquid to be sprayed, any attempt to color too fast will almost surely result in paint dripping down vertical surfaces. The gun must be kept moving to avoid this, to deposit only a slightly-merging mist with each pass, and waiting between passes, as

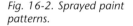

Fig. 16-2. Sprayed paint patterns.

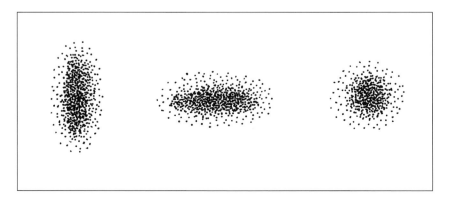

Fig. 16-3. Spray painting a car model. Photo: Yves Renaud. Design and model: Yves Renaud, Michael Pickstone and Isao Sasaki.

will be described later in this chapter. Only on the last coat should the gun slow down for the final shine. The trick is to get the last coat to merge very liquid and shine, leaving no drying bumps on the surface, but not enough liquid to cause dripping.

Drips can also be prevented by planning the spray work ahead of time. Inside (concave) curves, corners and edges are harder to reach with the paint mist than flats or outside curves. Even electrostatic spraying, used on aircraft, for example, doesn't reach inside areas very well. To spray difficult areas, the gun must be held closer, with the nozzle adjusted to a small pattern. Such areas should be sprayed before the easier, more accessible ones. If the opposite is attempted, the extra paint needed in the tight corner may wet the surrounding area too much and too soon after it has been coated, and the paint may drip.

Try to avoid "overspraying" past the model. As the spray gun is moved back and forth (or up and down) to spray, the movement should take it past the model at each pass, for a more uniform coat. Release the trigger of the gun each time it moves beyond the model. This will take a little practice, but it's worth it. Doing this avoids wasting paint and needlessly plugging up the spray booth filters. It also prevents some "dry spray" from flying around to another surface on the rear of the model—one which may already be sprayed with a good gloss coat. Some spray guns have dual-action triggers. Squeezed part way only compressed air is released. Squeezed all the way, the air-paint mixture is released. With these guns, instead of releasing the trigger after each pass, a partial release allows air to continue to flow. This helps dry the paint and blow away dry spray.

The spray gun's paint feed is usually adjustable. This allows more or less paint to be projected out with the compressed air. The result is a bigger or smaller pattern of paint deposited on the surface if the gun is not moving. The smaller pattern gives more control when spraying difficult areas and saves paint when spraying small surfaces.

The paint-air feed can also be adjusted, allowing more or less paint-air mix to be projected, but not affecting the size of the pattern. On easy work like large surfaces, a high feed will allow the work to go faster. On small work, a lower feed allows slower moving of the gun and more control. This feed can also be changed for various paint viscosities and for special effect spraying.

Referring again to the adjustment of the nozzle, the vertical pattern is most often used with a back and forth movement of the gun along the surface. The horizontal pattern is used with an up and down movement of the gun. The circular pattern, with the paint feed reduced, is used for small or difficult areas.

The size of the nozzle orifice may also be changed by using a different nozzle. This is usually done for different paint viscosities. A 1.2 to 1.4 mm. orifice is common for cellulose and acrylic lacquers, while a much bigger one, such as 2.5 mm., is needed for thick, viscous polyester fillers and heavy emulsions.

The paint viscosity can be controlled by varying the amount of solvent (thinner) in the mix. Viscosity not only affects the way the paint leaves the gun and flies through the air, but also how the droplets deposited will flow into a film on the surface before drying. Practicing with this variable can eventually be as good as following a manufacturer's recommendations or using viscosity measuring cups. Practice will also help you remember to keep the covers on cans of mixed paint. Solvents evaporate quickly from open cans, and the viscosity for the next spray job may increase significantly in a warm room in a short time.

The compressed air pressure is usually adjusted upstream from the gun, either with a pressure regulator at the spray booth connection or at the compressor. The air pressure needed is mostly related to the paint viscosity. Low pressure is used for milky liquids such as lacquers, and a higher pressure is needed to blow out viscous fillers and paints. There are also special effect sprays, which will be discussed below, that call for playing with the air pressure.

The temperature and humidity of the air in the spraying and drying areas influence both the initial flashing off of the solvents from the painted surface and the final drying time and finish. Temperatures of 45°–63°F (25°–35°C) are usual for almost all spraying, and even warmer for fast drying. In the spray booth, too low a temperature or too much humidity will often result not only in slow drying, but a surface haze called "blush." This can be polished off, however, if it remains on the last coat. Humidity in the compressed air can also cause problems. The water vapor in air quickly condenses to water when air is compressed and stored in tanks for spraying. These tanks should be bled regularly and the air lines opened long enough to blow out the water in them with the compressed air. This should be done more often in hot, humid weather, when water will quickly build up as it is compressed. Often, tiny pits that open on a painted surface, during or just after spraying, indicate that water droplets are being deposited with the paint.

The spray booth extracted air speed is not normally adjustable. This extraction serves a purpose so important that in countries like Switzerland, the law requires spray booth installations to have an automatic linkage between this ventilation and the compressed air. If no extractor ventilation is turned on, compressed air will not get through to the spray guns, which produce plenty of toxic vapors from solvents. This ventilation produces a wind past the person spraying and past the model being sprayed. It affects the paint drying speed, tendency to dry spray and dust landing on the sprayed surface. Probably the only way to effect a change in wind speed here is to increase the distance of the work from the spray booth filters. Shutting off the ventilators, when possible, or plugging or removing the filters can damage your lungs, the outside air, and/or the ventilator turbines.

The cleanliness of the surface is very important, as has already been pointed out. There is not only the problem of dust landing on a sticky surface. Nothing will hold well to a dusty, greasy or humid surface. The air carried into the spray booth can ruin work with not only dust particles, but with chemicals such as silicones and polyester resin vapors.

A completely dry model surface is required to hold paint for any length of time. Even if a model seems dry, it may have absorbed moisture during earlier work, and this moisture is almost surely going to try to evaporate from the surface by blistering and peeling the paint. Wiping the model clean of dust and grease with a damp rag might be a likely cause, so allow time for drying before painting. The most common cause of moisture problems, however, is wet sanding. One former student compounded his problems enormously by wetting his sandpaper in a bucket that had been used by someone else to mix silicone rubber. His model became one big "fish eye," refusing to accept paint even for a few minutes.

SPRAYING MULTIPLE COATS

As mentioned earlier in this chapter, it is not a good idea to try to color a surface by spraying one heavy coat. Multiple coats are needed to control the surface quality.

There are two basic approaches to spraying multiple coats. One is the "get it done now" spraying, and the other is the "let it dry overnight" approach. The first means staying with the model in the spray booth until the job is finished. A first, light coat is sprayed, resisting the temptation to see the model get colored right away. This prevents heavy paint dripping. Let this coat begin to dry in the wind created by the air being extracted through the filters. A boost can be given to the drying by lightly blowing compressed air on the painted surface. A few minutes later, much of the solvent will have evaporated and the surface will be tacky. This means two things: first, this tacky coat will accept another coat without dripping; second, this tacky coat will just begin to dissolve in the next coat's solvent before it, too, evaporates; this results in a good bond between the two coats, almost like one heavy coat that will not drip. Spray successive coats, waiting for each coat to partially dry before spraying the next. On the final coat, the gun can be slowed down a bit to give a heavier layer that will merge and shine, hopefully eliminating orange peel. If you don't dare go for that risky shine, the orange peel can be polished off once the paint is completely dry. This may take a few days if a lot of coats have been applied, especially if the air in the drying room is cold or humid.

The second approach is to let each coat dry completely before spraying the next one. This may be necessary. For example, someone else may need to use the spray booth, or an accident (dust, drips, etc.) may force you to discontinue the work. But this dried paint will not dissolve in the next coat to form a good "solvent" bond. It will be necessary to sand it lightly to reduce its gloss to a "keyed" surface that will hold the next coat. Sanding requires each coat to be perfectly dry to avoid smearing, so this approach can take as long, or even longer, than the drying time needed for the multiple coats in the first approach.

No matter how you spray all the coats needed, be prepared for possible slow drying and for the little accidents that might happen. Spray painting, as critical as it is after all the work in making the model, should never be planned for the last day or so. A "wet paint" sign, or no paint at all, will not help the model's presentation.

SPECIAL EFFECT PAINTING

Some special effects are obtained with brushing and rolling and with special paints, such as the wrinkle finish matt black used on aircraft control panels. The following techniques are employed with spray guns.

Many of the special effects discovered for use are just lucky finds. Playing and experimenting with pressures, paint mixes and distances will often lead to some exciting discoveries. You may one day want to use one of these effects for a part of your model, be it a car bumper or a camera button. Even if you never do, the exercise of trying and discovering all the "mistakes" that normally ruin gloss coats can be part of the fun of modelmaking. There are already some easy and effective special techniques available.

MATT FINISHES

Matt surfaces may be produced in several ways. Matting compounds may be bought to mix with gloss lacquers to obtain eggshell or matt finishes, depending on the amount of matt compound mixed in. These compounds are usually specific to the type of paint used, so it is important to buy compatible products, preferably of the same brand (a paint manufacturer won't usually bother testing another company's products to see if they all work together). It is also important to follow the dosing instructions on the label to get the desired effect. A few manufacturers produce a transparent matt compound that is compatible with their product, and that helps a lot. Often, the mix requires a high ratio of compound to paint, and once you have chosen the color of your dreams, you don't really need that gray or brown matt compound to kill your color. Another way to get matt finishes is to use the "problem" of dry spray to your advantage. Dry-sprayed surfaces are definitely matt—a velvet finish, different from what you can get with a matt compound. If the presentation of the model will allow it (this dry, chalky finish does not adhere well to surfaces), this may be just the matt you want. Hold the spray gun far away from the model while shooting, to allow the paint mist to dry before it hits the surface. Take your time. At that distance, not much paint will land on the model and the color will not cover fast. A big help is to color first, with normal-distance spraying. Then the far-off spray will only be to matt the surface, not to color it.

SPECKLE OR TEXTURE

It is easy to produce a speckle finish. The result is usually more effective if the speckle's color or shade is different from the base color. First, a desired color is sprayed, normally, on the surface. Then, using another color or shade of paint, put a thick mix into the spray gun. Somewhere between almost pure paint and half the normal amount of thinner is a good starting point for experimentation. As always, spray scrap surfaces until you get the effect you want for the model. The air

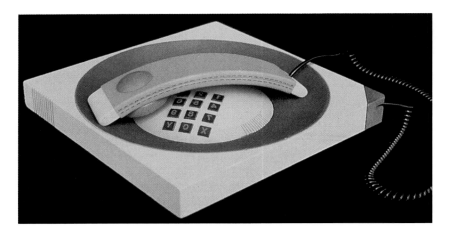

Fig. 16-4. A telephone model with a matt, slightly textured look. Photo, design and model: Ulrich Schraudolf.

pressure is reduced, and the gun's distance is increased a little. The nozzle size may be increased if the paint is too thick. All these variables will give different textures, visible as small drops, or speckles, on the surface.

CRACKLE

The crackle finish resembles old glazed pottery or antique lacquered furniture. One way to produce it is to use crackle finish paint. It is also just as easy to use matt compound to its "disadvantage" by mixing too much compound with the paint. The mix ratio is critical for matting. If you overdo the compound, the paint shrinks on the surface and the dried result is a crackle finish. The size of the crackles will depend on the overmix.

SPIDERWEB

The spiderweb finish looks just like its name, with thin strands of woven paint drying in a matt finish on the surface. Here, too, it may be more effective if the spiderweb spray contrasts a bit with the base color. To spray it, use pure, very thick paint. Some paints, even undiluted, are still not thick enough to produce this effect. Lower the air pressure, increase the gun distance and spray scraps to get the desired effect.

CLEARCOAT

Gloss transparent spray over the final color coat has its admirers. Car designers love it. One advantage of it is that it hides an orange peel surface, or sometimes even a poorly sanded and painted surface. It often hides too much, though, not only as varnish hides natural wood, but even with gloss lacquer colors, which have their own special shine. (It often ends up looking a little like candy apples.) The technique is simple: let the final color coat dry, then spray transparent gloss over it. Be careful to use a clear gloss that is compatible with the paint, in order to avoid ruining the final coat.

Fig. 16-5. Glossy finishes are a favorite of car designers. Photo, design and model: Alexandre Guye.

CARE OF THE SPRAY GUN AND MASK

Very little care is needed, either for the gun or the mask, but it must be regular and thorough. After each spray job, empty leftover paint into a labeled can and clean the gun. First spray used solvent through the nozzle for a few seconds, while wiping the exterior parts with solvent on a rag. Do not make the mistake of shooting solvent against the hand to wet and clean the outside of the nozzle. Blood tests on aircraft painters have shown that such a practice can inject paint into the body through the pores in the skin, and paint has sometimes been found halfway up the arm. Follow this initial cleaning by spraying a little clean solvent. To be sure that no paint will dry and clog the interior parts, the gun may be left in a solvent bath or in a closed container. The bath can be simply a wire mesh cage immersed in a container of used solvent, with the cover left on. Once in a while, dismantle the gun and check for worn parts and dried paint. The rubber joints may need to be replaced occasionally, and the few mechanical parts may need a small drop of oil. If the gun is used for polyesters, or any two-component paints or fillers, special, timely maintenance is needed. If you forget to clean it for longer than the pot life of the catalyzed resin, you may have to throw the gun away.

The use and care of the spray mask is very important, because it protects you from poisonous vapors. Wearing a dust mask is not enough, since this just filters dust particles from the air—and does a poor job of it at best. A "real" spray mask has a filter of activated charcoal. This adsorbs (as opposed to absorbs) molecules of many gases and vapors from the air as you breathe through it. Since almost all of the solvents used in modelmaking are toxic, this protects you, as long as the mask fits your face well and as long as the activated charcoal has not been saturated with poisons. Usually you can tell when the filter is finished, because you begin to smell the vapors, even through the filter. It is then time to replace the filter with new activated charcoal. When you're not spraying, storing the mask in a sealed container or plastic bag helps prolong the life of this charcoal. If left out in the air, it will soon adsorb enough pollution to be useless.

TROUBLE-SHOOTING GUIDE

The following is a quick-reference guide to the most common problems encountered in paint spraying, along with the causes and solutions:

- Specks and spots: It may be dust, and it may be anything or anyone— including you— causing this dust.
- Tiny craters: Water may be in the compressed air lines. Purge and blow the system dry.
- Fish eye: Tiny areas that repel paint as it is deposited are fish eyes. The cause is grease, wax or silicones on the surface. Some rub-on lettering is also too waxy to be painted. Use the heat-resistant variety.
- Orange peel: The paint is too viscous, the gun's distance is too great, the air temperature is too high, the gun is being moved too fast. There can be many causes of orange peel, and some professionals don't even try to correct it (just take a close look at a lot of new cars).
- Paint drips: The gun is too close or is moving too slowly. Try changing the position of the model to avoid vertical surfaces while spraying.
- Dry spray: The air pressure is too low, the gun is too far, the temperature too high, the paint is too thin, or you are overspraying past the surface. The spray booth filters may also be plugged up.
- Paint crackles when dry: Too much matt compound, or a wrong, incompatible solvent was added to the paint.
- Paint blisters when dry: The surface was wet or humid.
- Paint peels off when dry: The surface was either dusty, greasy or humid. Maybe the surface was sanded too smooth, or you forgot to prime the surface before painting.
- Bubbles and blisters show the original surface: The paint's solvent may have dissolved the (thermoplastic) primer. Use only compatible paint and primer systems.
- Slow drying paint: Many coats of paint will take some time to dry, especially in cold, humid air. The wrong (less volatile) solvent may have been used. The paint may be enamel.
- Paint spurting or leaking out of the gun: Dirty or worn spray gun parts may be the cause. The nozzle size may also be too big or too small for the paint's viscosity.
- Poor drying with blisters and wrinkles: There may be polyester resin vapors in the air, which can ruin thermoplastic paints.
- Model melts down or disappears: An unprotected thermoplastic model material may do just that.
- Painted watches or jewelry: If you hold your model to spray it, wear rubber gloves and leave your hardware at home.

LIST OF MODELMAKING MATERIALS

The following are lists of common materials used in design modelmaking. Each material's uses, advantages and limitations are briefly described.

SHEET MATERIALS

ABS Sheet: Easy to work. Often used in the United States for product design but hard to find in some countries.

Acrylic Sheet: Available clear and in many colors. Cast sheet costs more but is easier to solvent bond and optically better than extruded sheet. Hard to thermoform (bubbles form). Easy to paint (acrylic lacquer). Cuts with fine-tooth saw blades or repeated scoring.

Balsa Wood: Old-time toy modelmaker's wood. Very soft, light and easy to cut and shape.

Butyrate: Easy to cut by scoring and breaking. Not readily available in some countries.

Cardboard: Available in many qualities and thicknesses but not often used in modelmaking.

Cellulose Acetate (Clear): Easy to cut (up to 1.5 mm.) by scoring and breaking. Easy to thermoform. Bonds with acetone or methyl chloride.

Clear Polystyrene Sheet: Too brittle to be cut by scoring and breaking. Less expensive but optically poorer than acrylic. Easier to thermoform than acrylic.

Clear PVC Sheet: Brittle and hard to cut. Inexpensive and easy to thermoform, even with boiling water. Hard to paint, but polyurethane paints may hold.

Cloth, Rubber and Leather: Sometimes used when their natural surface texture is desired instead of trying difficult imitations.

Foam Core: Used for constructions with flat parts or simple bends. Easy to work but fragile and flammable. Useful for displays, architectural maquettes and all areas of industrial design.

Hardwoods and Fruitwoods: Difficult to work but give a great variety of excellent natural finishes.

Lime (Linden) Wood: The classical sculptor's and modelmaker's wood. Easy to cut and shape. A safe alternative to some synthetic materials.

Metal Sheet: Real aluminum, copper, lead, zinc, etc. may be the best choice for model parts that must resemble such materials.

Mirror Styrene Sheet: Inexpensive and fast for reflective surfaces, but not as good optically as real mirror.

Opaque PVC Sheet: Easier to cut than clear PVC. Easy to thermoform. Available in tubes and other forms as plumbing fixtures. Hard to paint.

Particle and Chip Board: Available in large sizes and many thicknesses. Not water resistant. Does not hold screws and nails well.

Plywood: Strong, dimensionally stable in humidity and now most often sold as "exterior" grade (has waterproof glue). Used for big constructions and armatures, or bucks, when strength is needed. Holds glue, nails and screws well.

Polycarbonate Sheet (Clear): Optically good. Resists impacts. Hard to thermoform (water content creates bubbles).

Polystyrene Sheet: For flat work or compound curves, by thermoforming. Inexpensive. Easy to cut by scoring and breaking. A thermoplastic soluble in acetone. Difficult to paint.

Pressed Fiber Board: Hard on saw blades. Not water resistant. Usually for temporary or hidden structures.

PVC Foam Sheet: Porous, matt surface. Hard to clean when smudged. Can be thermoformed.

SHAPING AND SCULPTING MATERIALS

Cast Solid Polyurethane: Hard and dimensionally stable but may warp with machining heat like wood does. Expensive. Non-porous and easy to paint. Accepts fine details.

Dry Bread and Foamed Starches: Not only bio-degradable but edible. Open-celled, requiring fast-drying fillers like plaster spackles.

Modelling Clays: Many types available for re-use or for permanent, hardened models.

Polystyrene Foam: Inexpensive and easy to shape. Various densities available for varying strength and detail. A thermoplastic, melting with heat and dissolving in many solvents.

Polyurethane (or Isocyanate) Foam: More expensive than styrene foam. Extremely dusty to work. Various densities available. A thermoset, which will not melt with heat or dissolve in solvents. Easy to shape, sand, fill and paint.

Pottery Clay: Another classical sculpting material, not often used for design models. Must be kept moist and wrapped in plastic between work periods. Armatures and hollow construction necessary for big models. May be fired for permanence or molded for cast reproductions.

Soap and Paraffin Wax: Safe, inexpensive and easy to shape. Can be cast, but shrinks. Cannot be painted.

Softwoods: Lime and balsa, available in block form. Easy to shape, sand, prime and paint. One of the safe alternatives to synthetics.

Solid and Foamed Plaster: Easy to shape when dried. Even more messy than clay. Safe to work. Holds paint if completely dry.

Styling Clay: Heavy, fragile, slow to work, smells awful and very hard to paint. Gives good detail and easy to work for add-on design changes. Used mostly in car styling.

CASTING AND MOLDING MATERIALS

Acrylic Casting Resin: Expensive and requires special equipment for casting. Gives high quality clear casts but seldom used for design models.

Cements: Fast-setting cement, dark fondu, pure white, etc. used for hard, permanent or special-effect casts.

Chocolate and Sugar Candy: Casting materials for temporary, edible models.

Clear Casting Polyester Resin: Low-viscosity completely colorless resins are the best. Requires a lot of care in casting. May be very toxic.

Dental Molding Plasters and Gels: Inexpensive, safe, flexible molding materials. Not always compatible with some casts (like high temperatures and polyester resins). Easy and fast. Some types more flexible than others.

Epoxy Resin: Low shrinkage. Used for some small casts. May be very bad for the health. Sets very hard.

Hot-Melt Glue: Can be used as a cheaper substitute for RTV for slightly flexible molds.

Metals: Low melting point metals, such as Wood's metal (melts in hot water), tin and pewter alloys may be used for easy, permanent casts. Some molds may melt with the heat.

Natural Gelatine: Safe and cheap for flexible molds, but can be difficult to work.

Plaster of Paris: Used for molds or casts. Fast (5 minute) setting. Gives good detail because plaster expands slightly on setting. Sets hard, requiring open angles as design limits. Open cells require sealing. A healthy, cheap, classical and messy material.

Polyester Resin: Used for big molds (reinforced with glass fibers). May be cast but shrinks and can burn if over-catalyzed. Sets very hard. May be very toxic.

Polyurethane Casting Resin: For fast (5 minutes) casts. Dimensionally stable when hard. Shrinks very little on setting. Gives good cast detail. May be very toxic.

Silicone (RTV) Rubber: Various flexibilities and hardnesses available. Expensive. Requires no sealer or separating agent for nonporous casts and master models. Not all types are compatible with some materials (like sulfur in styling clay). Very good for details and negative draft angles.

Urethane Foaming Resin: Two-component system expands on setting to produce a foam. May be very toxic. Limited storage time (due to humidity). Can be used for casts, molds or shaping solid foam.

VARIOUS LIQUIDS AND PASTES

Acetone: A solvent for some thermoplastic paints (like cellulose laquers). Dissolves styrene for solvent bonding. Common nail polish remover seems to be something else these days. Flammable and requires good ventilation.

Acrylic Lacquer: More flexible when dry than cellulose. Available as a solvent-based lacquer or water-based paint.

Alcohol: Wood or ferment distilled alcohol most often used. One of the first (and safest) solvents to try when cleaning a surface that may dissolve in other solvents.

Anti-Silicone Compound: Used to treat surfaces before painting if the presence of silicone is suspected.

Catalysts: Various chemicals that promote polymerization of polyester, epoxy, polyurethane, etc. resins. Dosage is critical for setting speed and quality of results. Poisonous.

Cellulose Lacquer: The classical lacquer. Dries fast and hard but becomes brittle with age.

Epoxy Resin: A very strong glue and a resistant paint. Does not dissolve styrene as polyester does. Poisonous no matter what the manufacturers say.

Etch Masking Tar: Used to mask metals for chemical etching.

Ferric Chloride: Used for copper etching. Poisonous.

Glues: Many types available for many materials and surfaces (porous and nonporous).

Kerosene (Petrol): Often used to protect tools from rust and to reduce friction on planer and circular saw tables. Will clean (dissolves) oil spots but is itself too oily to dry well.

Lacquer Thinners: Most are mixtures of solvents (toluene, xylene, acetone, etc.) for specific use on certain materials or for various drying speeds.

Linseed Oil: Made from plants. Perfect for natural wood finishes. Must be boiled or used with a drying agent for fast drying. Safer than synthetics in paints and putty.

Methyl Chloride: A "strong" solvent used in paint strippers. Bonds acrylic. Should not be breathed in.

Nitric Acid: Used for chemical etching of iron. Very dangerous to handle.

Paint Matting Compound: Available opaque or clear. Usually specific to paint types and brands. If overdosed it creates crackle finishes.

Polishing Liquids and Pastes: Available in various grades for rough and final polishing of metals, plastics and painted surfaces.

Polyester Paste Filler: A thick mixture of resin with dust (usually talcum powder). Easy to sand. May dissolve thermoplastic surfaces. May be toxic.

Polyester Resin: Used for gluing, filling (with powders mixed in), casting and molding, lay-up (with glass or carbon fibers) and gelcoat. Shrinks. Hard to sand. Very easy to paint. May be very toxic.

Polyester Spray Filler: A thin mixture of resin and dust for spray gun use. Easy to sand. Can be cast. Easy to paint even without primers. May be toxic.

Polyvinyl Alcohol (PVA): Used as a separating agent on porous (but sealed) molds. Very poisonous.

Primer Thinners: May be specific to certain primers only. Some contain chemicals, such as phosphoric acid (for aluminum) to etch surfaces for better adhesion.

Primers: Used on surfaces to fill pores and to increase paint adhesion. Many primers are specific to certain materials or some paints only.

Resin Colors: Available as opaque or transparent (for clear casts). Poisonous.

Shellac Base (Gum): Dissolves in alcohol to produce shellac. Inexpensive and easier to find in some countries than ready-mixed shellac.

Shellac: Dries to seal porous molds and to protect styrene foam, styling clay and other thermoplastics from other solvent treatments. A safe, natural material.

Silicone Spray: Such a good release agent that it is banned from many shops because its presence repels paint (fish eye) as it is applied.

Sodium Hydroxide: "Caustic soda." Used to etch aluminum. Very dangerous to handle.

Spackle Fillers: Plaster-based patching and filling compounds. Safe, inexpensive substitutes for synthetic fillers. Easy to sand and fast-setting. Do not attack model surfaces as polyesters may. Some types more flexible and adhere better than others.

Styrene Monomer: Better than solvents for thinning polyester resin since it reacts to become part of the resin, shrinking less. Very toxic.

Turpentine: Made from pine tree resin. Used to thin linseed oil and oil-based paints. Smells better and may be safer than its substitute, white spirits.

Two-Component Paint Systems: Polyester, epoxy and urethane resins. Thermosets. Polyurethane and epoxy also exist as solutuion-based or water-based paints. Dry fast and hard but may cause problems with spray guns. Poisonous.

Vaseline: Natural petroleum jelly. Used in thin coats as a release agent on porous molds and for thermoforming.

White Spirits: A synthetic substitute for turpentine. Smells worse but is less oily and evaporates faster.

VARIOUS MATERIALS

Found and Ready-Made Objects and Materials: Medicinal pills, wooden dowels, screen and wire mesh, hardware of all kinds. These are just a few examples of what may be used instead of making parts from scratch.

Masking Tape: The best is low-tack tape for safe removal from painted surfaces. If not available, reduce its tack by rubbing it on clothes or slightly dusty surfaces.

Piano Wire (and Spring Steel): Available as wire or thin, flat stock. High carbon content allows tempering for strong small tools specially made for some model work.

Silver Solder: Used for fairly low-temperature brazing strong joints on steel tools and for repairs.

Tack Rags: Sticky cloths used for final cleaning before painting.

Talcum Powder: Used in resin mixes for easily-sanded fillers. Used as a mold release in thermoforming.

INDEX